VOLATILE STATE

VOLATILE STATE

Iran in the Nuclear Age

David Oualaalou

Indiana University Press

This book is a publication of

Indiana University Press
Office of Scholarly Publishing
Herman B Wells Library 350
1320 East 10th Street
Bloomington, Indiana 47405 USA

iupress.indiana.edu

☉ The paper used in this publication meets the minimum
requirements of the American National Standard for Information
Sciences—Permanence of Paper for Printed Library Materials, ANSI
Z39.48-1992.

Manufactured in the United States of America

Cataloging information is available from the Library of Congress.

ISBN 978-0-253-02966-9 (cloth)
ISBN 978-0-253-03118-1 (paperback)
ISBN 978-0-253-03119-8 (ebook)

1 2 3 4 5   23 22 21 20 19 18

Contents

Preface

THE IDEA OF writing this book became more compelling after I completed my first one, *The Ambiguous Foreign Policy of the United States toward the Muslim World: More than a Handshake*. While researching scholarly material for my first book, I learned new perspectives on events evolving in the greater Middle East. Among the issues that captured my attention and interest is the United States' continuing negotiations with Iran over the latter's nuclear program. Research led me to question what will become of the Middle East when Iran becomes a nuclear state. The new insights and intellectual curiosity persuaded me to both express my thoughts and provide my own perspective on the future geopolitical landscape of the Middle East in postnuclear Iran.

This book, then, is about geopolitics and the future of international relations and US foreign policy toward both the Middle East and other major powers. A nuclear Iran in the Middle East, a region of always-dangerous flux and constantly shifting attitudes and energies on the global stage, will introduce new shifts in geopolitics and make international relations among major powers tenser. This result will affect not only the already tense relations among major world powers, mainly China, Russia, and the United States, but also among key regional players in the greater Middle East. A nuclear Iran introduces fundamental changes to a region where ambiguity, chaos, and violence are the orders of the day. In such a complex, often volatile region, relations between the United States and the region require careful attention to policy details—certainly more than diplomatic exchange in a new geopolitical landscape.

Before I write further about this topic and my insights on it, let me share my own background and how it qualifies me to write on geopolitics in the Middle East and postnuclear Iran. I worked in the field with elite military forces for over ten years carrying out operations at the behest of US policy makers. I speak Arabic and French fluently, understand the dynamics of the Middle East, grasp the depth of its social and cultural interactions, and know, from both abstract and practical perspectives, how the region's internal politics work. This firsthand experience is necessary, in my opinion, to offer the most salient analysis of how a nuclear Iran will change geopolitical calculations in the region among both major powers and regional countries.

Conventional wisdom suggests the new geopolitical landscape in the Middle East postnuclear Iran allows the latter to flex its military, economic, and ideological muscles with the support and help of Russia and China. Given these realities, Iran is closer than ever to realizing its political objective: becoming a regional power with the prestige and influence to guide and impact the region's political course. Against this backdrop, I argue that the United States' policy toward the region will have to adapt to these realities. Further, this policy will be tested like never before, especially when it comes to what kind of cooperation the United States undertakes with major powers in this new geopolitical environment. Will confrontation over energy resources in the Middle East trigger a military conflict? Alternatively, will the United States and other major powers, mainly China and Russia, follow a policy of appeasement to preserve peace and stability? Whatever the case may be, a nuclear Iran provides Russia and China the right platform on which to fortify their presence in the region and further challenge the much-declining American leadership and dominance of the last half century.

American foreign policy in the Middle East is marked by inconsistencies in its dealing with the region: the Iran-Iraq War, the invasion of Iraq, the civil war in Syria, the emergence of ISIS, the chaos in Yemen, and the political and security instability in Egypt, to name but a few. After all, American foreign policy affects the everyday lives and security of the inhabitants of Jordan, Iran, Turkey, Egypt, Iraq, Lebanon, Libya, Saudi Arabia, and Syria, among other nations. However, an intense rivalry among regional powers, mainly Iran and Saudi Arabia, marks the new postnuclear-Iran era. Further, other major powers will find it politically convenient ultimately to challenge the United States economically, militarily, diplomatically, and politically.

My writings include op-ed pieces on global affairs for the *Slovak Atlantic Council*, an affiliate of the *European Atlantic Council*, the *Waco Tribune-Herald*, *South China Morning Post*, and the *Huffington Post*. My articles have been translated into Arabic, Chinese, Farsi, French, German, Malay, Russian, Thai, Turkish, Spanish, and Vietnamese. In these articles, I address topics pertaining to international security, global affairs, relations among major powers, and ambiguous American foreign policy, among others. Occasionally, I describe what the geopolitical landscape in the Middle East might look like once Iran reaches an agreement with the United States over its nuclear program and the international community ends its economic sanctions and arms embargo on the Islamic republic.

This book is a work of scholarship, not an editorial; it informs not only people who keep up with current events but also students of American foreign policy, specifically those considering a career in international affairs. My analysis also draws from the writings of top scholars on the history, culture, and politics of Iran and the broader context of international relations, and from leading scholarship on American foreign policy in the region. The research herein is from top academic journals, books, and other scholarly writings. It should appeal to academics, foreign-policy professionals, and international affairs analysts. Although this work discusses unclassified or declassified documents, it does not and *cannot* offer factual evidence from classified sources.

My particular focus in this work attends to two important claims: (1) a nuclear Iran will inevitably spur nations like Turkey, Jordan, Egypt, the United Arab Emirates, and Saudi Arabia to pursue nuclear technology, thereby starting a nuclear arms race. Because of this geopolitical shift, major players will be forced to go back to the political drawing board to come up with new strategies; (2) the United States lacks the political will and vision to reconsider and reformulate its ill-conceived foreign policy in the Middle East and beyond. Suffice it to say, the ongoing turmoil in the region, which includes sectarian violence in Iraq, an ongoing civil war in Syria, upheaval in Yemen, the failed state of Libya, the social and political breakdown in Egypt, and the ongoing rift between Saudi Arabia and Iran, has not only heightened anxiety in the Middle East but also reinforced the notion of declining American leadership in the region. My concern lies in having China and Russia capitalize on US political weakness, obscure vision, and misaligned priorities while emboldening Iran through economic ventures, military upgrades, and diplomatic cover in the international forums.

We cannot address the new geopolitical landscape in the Middle East without addressing the impact the Islamic State of Iraq and ISIS are having on the political debate and psychological welfare of the region. Those developments suggest that postnuclear Iran will certainly differ from that to which we have become accustomed: new relations among major powers; a new economic, military, and ideological outlook for Iran; and new alliances. These new realities provide a framework that puts this major geopolitical shift in the context of international relations among major powers and between key regional players as events in the Middle East change by the hour, and ambiguity and chaos are the only constants. After reading this book, one will be in a position to answer the following questions:

- What kind of foreign policy does the United States intend to pursue for postnuclear Iran?
- With new geopolitical realities in the Middle East, how does the United States intend to engage other major powers, mainly Russia and China?
- Will competition over energy resources in the Middle East trigger a military conflict that could have China and Russia team up against the United States?

I write this book with the certainty that American foreign policy lacks clarity, vision, and a sense of direction. The United States must realize that it can no longer continue the status quo that it has held since World War II. A nuclear Iran and a shift in geopolitical priorities in the Middle East will introduce a new approach to international relations, new dynamics that require the United States to reformulate its foreign-policy strategy. As shown in the following work, the new geopolitical realities—the framework under which relations among major powers, including the United States, will be conducted and the economic, political, and military rise of China—make the task one only for the brave. Many around the world wonder, as I do, what the Middle East might look like post-nuclear Iran and whether the United States made a strategic mistake when it reached an agreement with the Islamic Republic of Iran concerning its nuclear program.

Washington must quit squabbling and focus on what lies ahead for post-nuclear Iran as regional balance of power shifts including expansion of Iran's influence and an Turkey's assertive foreign policy. The United States also needs to heed how China and Russia intend to behave in the region given this inevitable geopolitical shift. American policy makers assume that a direct conflict with China is unlikely. However, the possibility for such conflict exists if (a) China's economy continues its healthy growth; (b) the United States engages in another war in the Middle East that could interrupt China's access to energy sources there; and (c) China embarks on an assertive foreign policy in the Middle East in addition to the one in East Asia. With a high degree of certainty, I suggest that the national security interests of the United States and China will collide.

The United States must reconsider its global strategy given these force-producing events. One cannot assume constant, stable relations among major powers when their strategic interests are at stake. *Volatile State: Iran in the Nuclear Age* sets out to prove my point.

Acknowledgments

First, I would like to thank my wife, Michele, for her unwavering support, understanding, patience, and encouragement. She truly is my inspiration and motivation for writing more books. I also thank my wonderful boys, Layne, Beau, Michael, and Jaxon, for always being understanding. I'm sure one day it will make sense to them why I had to spend countless hours in front of the computer.

For his commitment, friendship, and help preparing this book, I extend my thanks to my editor, Dr. Les Ballard, with whom I worked extensively on this project. Les is an accomplished editor with great insights; the book is far better for his efforts. I would also like to thank the professional editorial facilitation of Ashley Runyon, acquisitions editor at Indiana University Press, and the entire staff. Ashley has truly made the process with this fine publisher effortless and delightful. I have been impressed by the publisher's care for the manuscript, confident decision making, and service to its readership. To Ashley, staff, and Indiana University Press at large, I say thank you for the opportunity to publish this book.

This project wouldn't have happened without the encouragement of my lifetime friends and supporters with whom I bounced countless ideas on global and geopolitical affairs, often over Friday-morning breakfasts. They were generous in sharing their time, wisdom, insights, and feedback with me. I do not mention them by name, as I will inevitably forget someone. You know who you are, and you know the debt of gratitude that I owe you. To all of you, I say thank you.

Finally, as a US Army veteran, I extend my thanks and appreciation to all service members of the US Armed Forces for their selfless service and patriotism. It has been an honor serving with you all.

VOLATILE STATE

1   Introduction

What will the geopolitical landscape of the Middle East look like in the coming years now that Western powers and Iran have reached a historic agreement over the latter's nuclear program? The answer is anyone's guess. Undoubtedly, this agreement ends Iran's regional isolation and further complicates regional politics. However, before discussing the nuclear accord the West reached with Iran, one must consider how this new agreement will impact international relations not only among major powers but also within the Middle East. The future geopolitical landscape of the Middle East will undoubtedly change. That landscape will be one in which Iran strengthens and flexes its military, economic, and ideological muscles with the support and assistance of Russia and China. The result is shifting loyalties and the emergence of new alliances in response to this new geopolitical shift, which will certainly have a major political impact on the ground. That is one of this book's key themes.

The Middle East is undergoing a major political transformation that compels regional and major powers, including China, Iran, Turkey, Saudi Arabia, Russia, and the United States, to go back to the drawing board to determine what strategies they should pursue to secure their respective interests in the region. Further, ongoing upheavals in the Middle East provide Iran with the right political platform to expand its sphere of influence. Consider, for example, the civil war in Syria that shows no signs of abating; anarchy in Yemen that already exacerbates tensions between Iran and Saudi Arabia; ongoing sectarian violence in Iraq that fuels tensions between Shiites and Sunnis across the region; political and security instability in Egypt; and a failed state in Libya.

We cannot address all the turmoil without highlighting an important fact: the emergence of the Islamic State in Iraq and al-Sham (hereinafter referred to as ISIS) may pave the way for Iran to demonstrate both its political influence and its ability to defeat ISIS. ISIS has come within close proximity to Iran's borders, and this threat to Iran has begun to redefine its geopolitical priorities in the region—it could even result in Iran and the United States agreeing to join forces against ISIS.

Now, the argument over the nuclear accord between the West and Iran is controversial. Some countries, mainly Israel and Saudi Arabia, have expressed their dissatisfaction with this agreement. Each country, whether in the Middle East or elsewhere, has its own political motives for supporting or undermining this agreement. Above all, the agreement, hoped for by so many, turns out to be contentious because of competing agendas. Even the United States Congress has already made clear its intention to reject the Iranian nuclear agreement. Now comes a political face-off between the legislative and executive branches of the American government. The much-anticipated quarrel regarding this agreement inside the Beltway shows the rest of the world the deep divisions in the US government.

This outcome leads many countries to question whether the United States can still play a global leadership role. I wager that Russia, China, Germany, and France will oppose any change to the agreement, since lifting sanctions against Iran provides an economic opportunity of great benefit. For instance, Germany has already dispatched a large delegation headed by the minister of trade, Rainer Brüderle, to Tehran to further his country's corporate interests in Iran. France's foreign affairs minister, Laurent Fabius, embarked with similar objectives.[1]

This nuclear agreement with Iran also benefits China, which eyes major economic ventures with Iran, mainly in the energy sector, after the lifting of sanctions. These developments suggest that this painstakingly negotiated agreement will have significant economic, military, and ideological effects at all levels. It could prove a game changer when it comes to geopolitics in the Middle East. I believe China, Russia, and some European countries will formulate their upcoming economic and diplomatic policies based on the political landscape of the Middle East. When formulating their policies, those nations will pay particular attention to how the emergence of Iran from isolation affects (a) global trade and (b) international relations. For instance, oil prices have already dropped in the international market.[2]

While addressing Iran's nuclear agreement and the inevitable shift in the geopolitical landscape of the Middle East, one sees that, whatever political changes take effect, the United States will still play a role in that system, no matter how limited. However, observers and global-affairs pundits question US ability to manage these force-producing dynamics given the latter's failed policies in the region, ranging from the invasion of Iraq and the war in Afghanistan to Syria's ongoing civil war and the conquests of ISIS. Add to that the failed state in Libya, turmoil in Yemen, and political and security instability in Egypt. Consequently, international security analysts ask whether these realities have forced the United States to cut a deal with Iran over the latter's nuclear program. There is a possibility. Yet, those who have argued against this agreement seem to forget that engaging in a military conflict with Iran will have far-reaching consequences that would certainly involve major powers. Margot Patterson, who has written extensively for various newspapers and magazines in the United States and abroad, is the author of *Islam Considered: A Christian Perspective*. She asks, "Yet is there any viable alternative to the deal? Are those arguing for a tougher agreement being realistic? Is a military strike or a war against Iran preferable? The answers are no, no and no."[3]

Israel and Saudi Arabia, to some degree, are behind the strong opposition of the US Congress to Iran's nuclear deal for two main reasons: (1) Israel worries that lifting sanctions against Iran paves the way for the latter to increase its military and economic support for groups such as Hezbollah and the Houthis, and (2) Saudi Arabia fears the expansion of an Iranian sphere of influence that may decrease Saudi Arabia's religious dominance in the Muslim world. However, I do not subscribe to the notion that Iran's postnuclear agreement advances either its ideology or, religiously speaking, its attempt to dominate the Muslim world. Even before the nuclear agreement, the invasion of Iraq, American foreign-policy double standards, the civil war in Syria, the upheaval in Yemen, the military coup in Egypt, and the failed state of Libya demonstrate that religious differences divide the Middle East in religious terms. In my opinion, Saudi Arabia is unwilling to relinquish its religious leadership.

I cannot fathom the logic of critics who argue that Iran wants to dominate the Muslim world. How could it, since the Shia population of about 170 million represents a fraction of the far larger Sunni Muslim population of 1.2 billion? Elsewhere I argue, "While Iran's political influence is undeniably growing, the Shiite camp does not wield influence or power when it comes to hard-core demographics. There are approximately 1.6 billion Muslims in the world representing 23 percent of the world's population. And of the 1.6 billion, only 10 percent is Shia. Even if Iran expands its influence in the Middle East, it will hardly have religious dominance over the Muslim world."[4] The Shia population is mainly limited to four countries: Iran, Pakistan, India, and Iraq. Limited numbers of Shia appear in Lebanon, Bahrain, Kuwait, Saudi Arabia, and Turkey, among others. However, what precipitates concern over sectarian tensions is not the agreement between the United States and Iran over the latter's nuclear program but the invasion of Iraq and the war in Afghanistan. Now, these tensions are more pronounced in countries like Lebanon, Saudi Arabia's eastern provinces (mainly Al-Qatif region), Iraq, and Yemen.[5]

We cannot overlook how the United States has helped pave the way for Iran to play a greater role in the Middle East. Of interest is the indecisive victory of US military actions in Iraq and Afghanistan. Each war eliminated one of Iran's chief regional enemies: Saddam Hussein in Iraq and the Taliban in Afghanistan. This political outcome has cleared the way for Iran to pursue its geopolitical aspirations. Now that Iran has reached this historic agreement with the West over its nuclear program, it is well positioned to exert more influence in the region, backed by economic and military apparatus. For instance, Iran's support for the al-Assad regime in Syria and the Houthi rebels in Yemen suggest Iran's desire to influence the outcome of both conflicts in favor of its long-term strategic interests.

The failure of US foreign policy in the Middle East is attributed to an ambiguous policy and ill-defined strategy. Iran understood early on that a chaotic

Middle East not only entangles the United States and drains its resources but also provides Iran with the opportunity to further its aspirations in the region. And that is exactly what happened during the invasion of Iraq, where the United States failed to understand that supporting a Shia government in Baghdad under the poor leadership of the former prime minister, Nuri Kamal al-Maliki, created a back door for Iran to manage Iraq's internal affairs. I argue that the policies of the current Iraqi prime minister, Haider al-Abadi, which consist of targeting and marginalizing Sunnis, are the same policies pursued by his predecessor, al-Maliki. The outcome is that a flood of Shiite volunteers have joined the fight, whether against ISIS or the Sunni tribes, and now reinforce the ranks of Shiite militias allied with Iran. Equally important, Iraq's purchase of Iranian weapons suggests close ties between both governments. Lolita Baldor argues, "Over the past year, Iran sold Iraq nearly $10 billion worth of weapons and hardware, mostly weapons for urban warfare like assault rifles, heavy machine-guns and rocket launchers, an Iraqi government official said. The daily stream of Iranian cargo planes bringing weapons to Baghdad was confirmed at a news conference by a former Shiite militia leader, Jamal Jaafar."[6] This close relationship between Iran and Iraq results from the ill-conceived US policy of invading Iraq in 2003. Some may disagree, but, however one judges the arms sale from Iran, it is nonetheless true that events including the Iraq-Iran war, civil war in Lebanon, Persian Gulf War, and the invasion of Iraq in 2003 have helped Iran to provide support to interested parties to execute its strategic objective: expansion of its sphere of influence in the Middle East.

History shows that certain conditions indeed preceded, and provided an opportunity for, the reemergence and revival of Iran's Shiite ideology; chief among them is the 1979 Iranian revolution. If the Iraq invasion revived Shiism after its long-dormant era, the 1979 revolution laid the foundation on which the ideology could withstand the test of time. For this reason, I assert, Iran strengthens its ideological ties and maintains its support to Shiite communities in Lebanon, Syria, Iraq, and Yemen. The objective has always been, and continues to be, for Tehran to enhance its political influence and expand its religious ideology. A strong military apparatus helps Iran project regional power. For instance, one should perceive the size of Iran's armed forces, which exceed a half-million personnel in active service, as part of its expansionist strategy in the region. Equally important, Iran's missile technologies—including the Shahab, Ghadr, and Sejjil systems, which can reach distances of 2,000, 1,800, and 2,000 kilometers, respectively—exhibit its military's ability to embark on any ventures in support of its aspirations for regional influence and dominance.[7] Yet, this has been in play since the Persian Empire. As Michael Totten states, "Iran has been a regional power since the time of the Persian Empire, and its Islamic leaders have played an entirely pernicious role in the Middle East since they seized power from Mohammad Shah Reza Pahlavi in 1979, stormed the US Embassy in Tehran, and held 66 American diplomats hostage for 444 days."[8]

China's Geostrategic Calculations in the Region

While addressing these developments, one also should emphasize the role that other major powers, like China and Russia, play in the new geopolitical landscape of the Middle East. In this section, I address each country's role in this new political landscape separately. Further, I focus my argument on how Iran's new status helps China and Russia to further advance their corporate interests and also challenge the declining regional leadership of the United States. I keep returning to the question of whether these interests, mainly on the part of China, divert the United States' focus away from the Pacific region given China's military buildup in the disputed islands in the South China Sea. In the subsequent sections, the answer sheds light on China's long-term strategy. For now, suffice it to say, China and Russia's interests in the Middle East will grow stronger, bolder, and more calculating, signaling a far more assertive foreign policy from both countries in the coming years.

Regarding China, observers have noticed in recent years its increased economic, military, and political footprints in the Middle East. This pursuit stems from two main factors: (1) China's need for access to energy sources mainly in Saudi Arabia, Iran—given its proximity to the Caspian Sea—and Turkey; and (2) China's economic involvement in the Middle East facilitates expansion of its political sphere of influence, thus projecting to the region its ability to challenge American leadership in the region. One may fairly state that the West, including the United States, did not expect China to be where it is today economically, politically, and diplomatically. Former US treasurer Henry Paulson states, "Forty years ago most Americans wouldn't have imagined owing China one red cent. Now it is the US's biggest creditor, owing just under $1.3 trillion of our government's debt. It's enough to make the head spin—or for Americans to wonder how the world got turned upside down so fast."[9]

China's economic preeminence presents major challenges to the United States not only in the Middle East but also in other parts of the world. In my opinion, dissatisfaction with failed, hypocritical US policies in the Middle East intensifies this challenge. Thus, for instance, key regional players like Saudi Arabia recognize China as a possible alternative to American hegemony there. This idea has become more attractive given the United States–Iran accord over the latter's nuclear program. Could this explain why China is developing major partnership ties with countries like Iran, Turkey, and Saudi Arabia? China's pivot westward is no more evident than during the meeting between the late Saudi king Abdullah and China's leader, Xi Jinping. The United States now tries to minimize its footprints in the Middle East after its disastrous invasion of Iraq and political paralysis in dealing with the civil war in Syria and Yemen. Meanwhile, China realizes that it makes sense to pursue its foreign policy in the region. China's policy would not challenge US leadership in the region, or whatever is left of it. Instead, China focuses on strengthening its economic ties in the region, mainly in the energy sector. Yun Sun, an author and nonresident fellow at the Brookings Institute, argues, "The underlying 'retreat-pursuit' philosophy of 'March West' still anchors on a zero-sum perception of the U.S.-China relations. As the U.S. pivots away from the region and becomes deeply immersed in Asia Pacific, it needs to seriously consider the implications of and prepare for China's 'March West.'"[10]

This approach suggests that China's long-term strategy focuses on having a relationship with the United States more complementary than acrimonious when it comes to the Middle East. Embarking on this policy, China achieves two objectives: (1) US attention shifts from the Pacific region, and (2) China secures more economic ventures with key regional players, including Iran, Saudi Arabia, and Turkey. Will the Middle East become a hot zone where the United States' and China's interests collide? The answer is possibly. That is because the lifting of sanctions on Iran following its nuclear agreement with the West compels China to move quickly toward securing economic contracts mainly in the energy sectors. Moreover, China and Iran may enter into a military treaty of some sort, clearing the way for the former to establish military and naval bases on Iranian soil, similar to the arrangement the United States has with Bahrain. However, that does not strike me as a wise policy for China to embark on at this time. Rather, China should wait until the dust settles in this latest major shift in regional geopolitics to better assess the situation and execute its next calculated move.

Yet some scholars, notably Charles L. Glaser, argue that all the United States wants to accomplish through its ill-defined policies, whether in Asia or the Middle East, is to maintain the status quo.[11] While China steadily increases its footprints in the Middle East, its rapprochement with Iran provides it a much-needed platform from which to pursue its strategic interests. Yet, I ask, how would Iran respond to the turmoil in the Middle East, including civil war in Syria with no end in sight, sectarian violence in Iraq, Libya's failed state, and upheaval in Yemen? The answer depends on many factors that, in my opinion, will impact relationships among not only major powers but also regional actors.

For instance, the ongoing military conflict in Yemen between Iran's backed Shia group, the Houthi rebels, and the Saudi-backed government led by Abd-Rabbuh Mansour Hadi suggests the conflict has wider implications that could impact the entire region. Iran's military and financial support for the Houthi rebels raises major concerns within Sunni countries, mainly in the Gulf region. Saudi Arabia formed a coalition of ten Arab nations to address the conflict, quickly convincing other Sunni countries that the threat of Iran is real; thus,

Saudi Arabia has argued that the need for contributing troops is paramount. Iran now tries to break the Saudi coalition. These happenings convince security analysts that the conflict in Yemen is much deeper than it appears on the surface—certainly more significant than many Western media outlets indicate in their fleeting news coverage. However, I do not go as far as some analysts who suggest the conflict could escalate into World War III.[12]

The history of conflicts in the Middle East (at least when it doesn't involve Israel) has always depicted the two main religious denominations within the faith of Islam: the Sunnis and the Shiites. To better understand the issues at the core of this eternal ideological conflict, one ought to have a basic knowledge of what Saudi Arabia and Iran represent, religiously speaking. On one hand, Saudi Arabia represents an ascetic, ultraconservative, and somehow twisted interpretation of Islam. On the other, Iran subscribes to similar religious beliefs except that Shiites believe in some practices uncommon among Sunnis. At the heart of the matter is not whether Shiites' religious practices and rituals differ from those of Sunnis; rather, the difference lies in the political and spiritual grounds of those beliefs.

The ongoing regional conflicts reflect the ideological attraction Iran is gaining in countries like Yemen, Syria, Lebanon, Bahrain, and other Gulf states. Moreover, even though two crucial Islamic holy sites—Mecca and Medina—are located in Saudi Arabia, Iran (and Shiites, for that matter) doesn't grant the kingdom pivotal leadership over the entire Muslim world. The two branches of the faith do not see eye to eye. I address this topic in great depth in the subsequent chapters and highlight Saudi Arabia's fear of Iran's emerging power, especially in the aftermath of Iran's nuclear deal with the West. Suffice it to say that the ongoing conflict in Yemen and Syria, for that matter, betrays US leadership decline in a chaotic Middle East. I argue that Iran, Russia, and China are hedging their bets that this trend will continue for months, if not years, to come.

Against this backdrop, the American foreign-policy establishment lacks the vision, strategy, and creativity to come up with pragmatic alternatives to its failed policies in the region since the terrorist attacks of September 2001. Could it be that these force-producing events in the region prove too much for the United States to handle? Critics suggest that is the case.

Similarly, a tense scenario presents challenges to the United States in the Pacific realm. Of interest is China's ongoing buildup in much-disputed territory in the South China Sea. China's ability to build an airstrip big enough for military aircraft proves, once again, US political paralysis and inability to influence events in the region.[13] The United States has no choice but to accept the new reality: China is displaying its military strength as part of a changing geopolitical landscape in the region. However, Charles L. Glaser's suggestion—to bargain with China—disadvantages the United States by permitting China and others to perceive the irrelevance of the United States. Glaser suggests the following: "A possibility designed to provide the benefits of accommodation while reducing its risks is a grand bargain in which the United States ends its commitment to defend Taiwan and, in turn, China peacefully resolves its maritime disputes in the South China and East China Seas and officially accepts the United States' long-term military security role in East Asia."[14] Embarking on this strategy sends a message to neighboring countries—Japan, Brunei, the Philippines, Taiwan, Malaysia, and Vietnam—that the United States' days of influence over the Pacific are numbered.

Against this backdrop, I do not see how China, knowing that it is succeeding in squeezing the United States out of Asia, would compromise its interests in the Middle East in the aftermath of Iran's nuclear agreement under the banner of cooperation with the United States. These political maneuverings suggest that, given its long historical ties to the region, China wants to seize the opportunity to reengage in the Middle East. In fact, those ties go back to 138 BC, when China's Han dynasty dispatched emissaries to establish economic and political relations with the region. China's "Silk Road" policy is a sensible model for how China now intends to pursue its strategy in the Middle East. This strategy is aided by none other than the historical agreements between Iran and the United States. As a result, Iran would be the platform on which China will eventually expand its economic strategies, which, in turn, make possible its military ventures. China's economic engagement with the region for more than a millennium allows it to link by land to Persia, making possible more maritime routes to the greater Middle East.[15] To illustrate its commitment to developing its economic infrastructure, China has invested an estimated $120 billion in contracts with the Iranian hydrocarbon sector.[16] In my opinion, this investment highlights China's ongoing desire to access energy sources in the region. I believe China could engage militarily if a hostile force interrupts the flow of oil that supports its economic machine. This explains why China is simultaneously engaging two major oil producers in the Middle East—Iran and Saudi Arabia.

The shifting geopolitical landscape of the Middle East compels key regional players, including Saudi Arabia, to reorient their focus toward China rather than the United States. There are two key factors to this sudden shift: (1) Saudi Arabia realizes that oil production in the United States is going up, resulting in less demand for oil from the kingdom; and (2) the shift of wealth from the West to the East following China's rise to economic preeminence presents huge economic opportunities for the desert kingdom to benefit from. More recently, China hosted a signing ceremony in Beijing that launched the China-led Asian Infrastructure Investment Bank. The bank will have about a $100 billion capital base in which China provides $30 billion of its capital. More countries, including Turkey, South Korea, and others, have expressed interest in joining the institution.[17] One cannot mistake the factors that contribute to this shift: China's long-term vision and its ability to create and rigidly pursue economic, military, and financial strategies over the past few decades have led to this strength.[18]

To put this in a geopolitical context, the lifting of sanctions against Iran provides incentives for both Iran and China to shape the political narrative in the region despite each country pursuing its own objectives. For Iran, reaching an agreement with the West unshackles Tehran's aspirations, supported by billions of dollars, to expand its sphere of influence and impact the region militarily, economically, and ideologically. China, on the other hand, will convince the world that its economic preeminence is not random but based on a systematic strategy, poetic calculations, fiscal discipline, and decades-long economic foresight. Further, China contributing to the geopolitical shift in the Middle East through its support of the Iran nuclear agreement sends a message that the rest of the world must treat China as a partner on an equal footing with the United States. Thus, it behooves the United States not to antagonize China. Doing so would not be in the United States' interest, especially in the Middle East.

Militarily, China's and Iran's interests could merge in an unexpected way. One assumes the lifting of sanctions on Iran will allow China to sell its advanced military systems to many countries in the region. After the international community lifts sanctions against Iran, it will be in a position to acquire Chinese military hardware on a larger scale. In turn, that may compel Saudi Arabia to do the same just to undermine Iran. To put the latter assertion within the context of the geopolitical shift in the Middle East, China's latest successful test of its hypersonic nuclear Wu-14 missile (more than 7,600 miles per hour)[19] has drawn the attention of many potential customers in the Middle East who have already started to see China as an alternative to the United States as a supplier of defense weapons systems.

Despite the much-anticipated geopolitical shift in the Middle East following the Iran-US nuclear agreement, I am convinced that China will proceed cautiously so as not to upset the balance of the world order, a system that allows it to acquire wealth, power, influence, and status. China strikes me as a pragmatic regime that understands the stakes if it flaunts its economic and military strength beyond what is realistic. But, make no mistake, China's massive buildup in the South China Sea, the introduction of its yuan as the only currency used for oil trade in Asia, the establishment of the Asian Infrastructure Investment Bank, and the security treaty with Russia, Pakistan, and India[20] send a clear message to the West that China will challenge the world order if its strategic interests are threatened. I believe China's foreign-policy decisions post Iran-US nuclear deal reflect a long-term strategy rather than aim at short-term gains.

In the subsequent pages, I address in great depth the role China will play in the new geopolitical landscape of the Middle East. Many international security analysts, scholars, military commanders, government officials, and pundits believe that role will have a major impact on international relations not only between the West and key Middle Eastern countries, including Iran, but also among world powers as they compete for global influence, even dominance.

Russia's Geostrategic Calculations in the Middle East

One cannot address the geopolitics of the Middle East without emphasizing the role Russia continues to play in the ongoing turmoil there. For instance, Russia's impact on the civil war in Syria is what hinders, to a degree, the West's ability to find a solution to the conflict. On different occasions, Russia has vetoed any international military intervention in Syria, thus ensuring that the al-Assad regime stays in power. This, in turn, allows Russia to further its interests not only in Syria but also in the greater Middle East.

Before delving deeper into this theme, it is imperative to note that the collapse of the Soviet Union in 1991 changed the world geopolitically, reshaping it from a multipolar system represented by the United States and the Soviet Union to a unipolar system in which the United States is the sole power. However, almost twenty-five years after the collapse of the USSR, Russian president Vladimir Putin now tries to reaffirm that Russia acts on the world stage as an equal to the United States. Hence, talk over the reemergence of a multipolar system has become the topic du jour within political circles. A difference between the balance of power before 1991 and today is that the new system is not limited to Russia and the United States but includes China, India, Brazil, and Iran.

International security analysts argue that the agreement between the P5+1 (the five permanent members of the United Nations Security Council—United States, France, China, Russia, and Britain—plus Germany) and Iran over the latter's nuclear program has provided Russia with a political and economic platform on which it can further its geopolitical agenda in the region. In support of the latter assertion, I argue that Russia's role in blocking additional sanctions on Iran before reaching the historic accord suggests that Russia knew that it could influence the outcome of the negotiations in its favor. That suggests that Iran and Russia were working together to circumvent US sanctions. According to Stratfor, "A multibillion-dollar trade deal between Russia and Iran to work around U.S.-led sanctions would be market-shaking news indeed and would place U.S.-Iran nuclear negotiations in jeopardy. But this is far from what is actually taking place."[21] Similarly, the United States knew during all stages of the negotiations with Iran over its nuclear program that it would be impossible to reach an agreement without Russia's support. Given these force-producing dynamics, would it be accurate to state that the world order operates under a multipolar system?

As stated earlier, the collapse of the Soviet Union in 1991 redirected the West's interest into expanding its sphere of influence in areas that were once under Russia's control. Of interest are the Baltic states of Latvia, Lithuania, and Estonia. Russia saw this expansion by the North Atlantic Treaty Organization

(NATO) as a threat to its security. Worse yet, the United States' decision to install an antimissile defense shield in Poland forced Russia to embark on strategies that it would not have pursued otherwise; consider, for example, Russia's annexation of Crimea. To gain Russia's cooperation on other issues, including the war in Afghanistan and nuclear negotiations with Iran, the United States has decided to scale back the final phase of installing the missile defense system. Russia's fierce opposition underscores its ability to influence outcomes of great importance to its geopolitical interests. David M. Herszenhorn and Michael R. Gordon argue, "The Obama administration has sought cooperation from Russia on numerous issues, with varying degrees of success. Russia generally has supported the NATO-led military effort in Afghanistan and has helped to restrict Iran's nuclear program by supporting economic sanctions. But the two countries have been deeply at odds over the war in Syria, and over human rights issues in Russia."[22]

The current political and security landscape in the Middle East suggests the two main security threats to the United States are Iran and ISIS. However, Russia—and China, for that matter—will prove to play a far greater role in the new geopolitical landscape in the Middle East than what is now being advocated. Russia's unpredictability makes this ever more pertinent. One recalls the speed with which Russia annexed Crimea and how that led to tensions between the United States and Russia not seen since the days of the Cold War. Russia's sudden use of force has raised questions in the West about whether Russia can play a constructive role after the Iranian nuclear agreement or if it will instead be a pariah that uses its position within the international system, as a permanent member of the United Nations Security Council, to undermine security and stability around the world. Russia has used its veto power in the Security Council to prevent international intervention in the civil war in Syria and to intensify sanctions against Iran. Those votes offer evidence of Russia thwarting the West's recent efforts to exercise influence in the Middle East.

What sort of foreign policy will Russia pursue in the Middle East given the nuclear agreement between Iran and the West? Will it pursue a confrontational foreign policy or seek a peaceful existence in the much-anticipated shift in geopolitical landscape in the Middle East? Answering these questions within the context of a multipolar system provides a better understanding of what to expect from Russia in the era after Iran's nuclear agreement. For now, Russia thinks that economic and military opportunities will precede this new shift in the political landscape of the Middle East before it defines its long-term strategies. In the subsequent pages, I address in great depth what this shift entails and how Russia will position itself to gain from this geopolitical change.

One must note that Russia's assertive foreign policy in the Middle East is not new. Recent Russian deeds show the thinking within Russia's elites that, though it may have lost the influence it once had under the Soviet flag, Russia intends to reassert itself to ensure a seat at the table with other international powers. For instance, Russia's declaration about fighting ISIS masks the real motive behind the strategy: ensuring that al-Assad stays in power. I argue that Russia's geopolitical strategy comes not only through Syria—assuming that al-Assad stays in power—but also through Iran. Could this explain why, in order to bolster its presence in the region, Russia forged an alliance with Iraq, Iran, Syria, and Hezbollah?

Meanwhile, the US government echoes, as usual, the empty rhetoric of how concerned our leaders are. While these statements obviously target our more ill-informed citizens, Russia moves forward with its strategy of purging Syrian antigovernment rebels while ensuring that al-Assad remains in power. Putin fully understands that it's now or never for Russia to reassert itself on the global stage. The upheavals in the Middle East, including civil war in Syria and Yemen, a failed state in Libya, and ISIS, among others, provide Putin the perfect opportunity to achieve his long-term strategic geopolitical vision.

Global-affairs analysts agree on one assertion: the Middle East has not reached the end of its share of conflicts. However, these prolonged conflicts provide Russia a raison d'être through its presence in the region and allows it to adjust its political tactics to influence the outcome to favor its geopolitical aspirations. One need not look far to realize how wrong some Western analysts were to suggest that the collapse of the former Soviet Union meant that Russia would never again emerge as a formidable power.

Yet, I ask whether the United States' behavior in the region—its involvement in unnecessary conflicts in addition to its ill-conceived foreign-policy strategy— paves the way for the reemergence of Russia (and China, for that matter). In support of the latter assertion, I argue that many in the West were astonished to learn that Russia and China are selling advanced missile technology to a major US ally, the Kingdom of Saudi Arabia. Further, the military and diplomatic support Russia and China provide Iran in international forums, including the UN Security Council, only strengthens my argument. Do you recall how Russia and China vehemently vetoed propositions against any additional sanctions on Iran? These dynamics unequivocally demonstrate that US leadership in the Middle East is indeed declining.

Equally important, consider Russia's vehement support for Iran during the latter's nuclear negotiations with the West, resulting in 2015's agreement. During all phases of the negotiations, Russia opposed any additional sanctions on Iran. Given the ongoing political tensions between Russia and the United States, the thinking in Moscow was that it would be in a position to compel other major powers, mainly the United States, to agree with Iran on its nuclear program. That is exactly what took place. What the US government did not disclose, however, was that Russia's efforts behind the scenes made it possible to reach a deal with Iran. As a result, the agreement bolstered Russia's status, providing Putin with a sense of pride and the possibility of influencing other major conflicts, including those in Syria, Yemen, and Libya.

These force-producing events suggest that Russia wants the West to acknowledge and not dismiss its security, economic, and strategic interests in the Middle East. I argue that the ongoing civil war in Syria, and Russia's direct military involvement there, highlights Moscow's long-term objective in the region: exerting more political influence. I even go further and state that, by taking a military risk in the region, Russia is sending a message to the United States that it will challenge its hegemony in the region. Russia intends the other message for NATO, following NATO's recent invitation for Montenegro to join the alliance. Russia perceives this move as a threat to its security. That is why, in my opinion, Putin decided early on to deploy advanced missile systems and heavy weapons to Syria. Given the current dynamics on the ground and the shifting geopolitical landscape, I believe Russia is communicating to NATO that it will retaliate in some form should the latter decide to expand eastward.

One could argue that Russia's concerns regarding the ongoing turmoil in the Middle East could extend beyond the region and reach Russia's North Caucasus region, where its Muslim minority resides. This religious and cultural dynamic could explain why Russia adamantly pursues an assertive foreign policy in the Middle East. Iran's latest nuclear agreement with the West represents the perfect opportunity for Russia to reshape its international political presence. As Stephen Blank observes, "Geopolitically, the black hole of Central Asia now constitutes an expanded part of the new Middle East. Geoculturally, few other regions entail a nation-state border system of such potential transparency, where common and cross-border religious, ethnic, linguistic, and collective memories could act individually or jointly as destabilizing or integrating factors."[23]

As it stands, Putin now capitalizes on the upheaval the Middle East continues to experience: anarchy in Yemen, conquest by ISIS, civil war in Syria, the failed state in Libya, and security and political unrest in Egypt, among other issues. Furthermore, Russia understands that, without its involvement, there can be no diplomatic solution to the Syrian conflict. At the same time, Putin understands that Russia's strategic interests in the Middle East are linked to keeping Bashar al-Assad in power. Putin's strategic calculations go beyond the borders of Syria. As support for my argument, consider that Russia has recently deployed some advanced weapons to the Syrian theater under the pretext of fighting ISIS. These weapons include fifty-five fighter jets, including SU-30, SU-24, and SU-25 models in addition to SU-34 bombers, along with seven Mi-24 and five Mi-8 helicopters. The base where these weapons are housed is also fortified with Pantsir-S1, Buk-M2, S-200, Pechora-2M, and S-400 air-defense batteries.[24] Further, consider the conflict in Yemen, which is basically a proxy war between Iran and Saudi Arabia. Russia provides Iran with advanced weapons that, at some point, could be funneled to Houthi rebels, a Shia group backed by Iran.

This comes on the heels of Saudi Arabia's latest pivot toward Moscow, providing Putin another opportunity to further his agenda in the region. However, the most important aspect of the Middle East that allows Russia to play an even greater role is the latest agreement between Iran and the West, an accord in which Russia played a major role behind the scenes to ensure the consideration of its interests.

Assessing Russia's long-term strategy in the Middle East, I argue that its approach, in concert with China, is to counter US dominance in the region. Both Russia and China—besides other regional players—are well aware of US leadership decline in the region, a decline attributed to an ambiguous foreign policy, double standards, and lack of vision. History shows the political turbulence that followed the collapse of the Soviet Union as the United States failed to lead the international community on its own. Consider, for example, the genocide in Rwanda, in which a paralyzed United States failed to intervene. Note the failed state in Somalia, which demonstrated US inability to engage in guerilla warfare. Observe North Korea's ability to acquire nuclear weapons, which showed a lack of US political influence over nations that could proliferate nuclear technology. Reckon the 9/11 attacks, which brought about a militarized foreign policy. Witness the invasion of Iraq, which demonstrated a lack of credibility in the United States. Contemplate the rise of China, which underlines the shift of wealth from the West to the East. Mark the civil war in Syria, which demonstrated the United States' ill-defined and weak foreign policy. Lastly, behold the conquests of ISIS, which show a lack of US understanding that the ISIS dilemma requires an ideological solution rather than a military approach. After all, ISIS demonstrates, thus far, that it is a political force with a political agenda wrapped in a religious narrative.

Against this backdrop, Russia's increased economic, military, and diplomatic footmark in the Middle East could feed into the narrative about US global leadership decline. Because of this perception, I believe the Iran nuclear agreement will eventually reintroduce the notion of a multipolar system. One asks if this could be the reason Russia and China provide Iran with diplomatic cover in international forums. Changes in the political landscape of the Middle East in the next five years or so would certainly provide a clear answer to this question.

In the following pages, I focus the discussion on how, despite having mutual interests, Iran and Russia aim to achieve separate objectives. While it's no secret that Russia maintains economic and military ties with Iran, its objective is long term. In my opinion, that objective is for Russia to establish a naval base in Iran in a similar arrangement to that of the United States in Bahrain. Doing so, Russia positions itself to challenge the United States economically and militarily more openly than ever before.

The ongoing upheavals in the Middle East—civil war in Syria, turmoil in Yemen, the failed state in Libya, political and security insatiability in Egypt, and ongoing sectarian violence in Iraq—have facilitated Russia's reentry into Middle Eastern politics. The nuclear agreement between Iran and the West will provide

Russia with even greater economic and, eventually, military opportunities to cement its presence in the region. However, in order for these opportunities to materialize, Russia believes that the regional conflicts will persist, allowing it to sustain and manage its presence in a chaotic region. Putin's strategic objectives are twofold: (1) the Kremlin (and, specifically, Putin) desires to regain its Soviet-era influence, and (2) since Russia played an active role in Iran's nuclear negotiations, which led to the signed agreement, Russia expects to assert itself as a major global player that wields power and influence. I unequivocally state that Russia's interests in the Middle East stem from its understanding of the strategic importance of the region.

With the anticipated shift in the geopolitical landscape of the Middle East, I predict that Russia's foreign policy in the region will remain steady as long as (a) the upheavals in the Middle East persist; (b) tensions with the United States over Crimea remain, providing Russia an excuse for not cooperating with the United States; and (c) Iran and Russia are able to have some sort of an agreement on regional matters (i.e., continued support for the al-Assad regime). These dynamics coincide with the July 2015 meeting between Russian and Iranian foreign ministers, adding to the speculation of much closer relations between the two countries. This rapprochement, in the works for quite some time, raises concerns within the Washington foreign-policy establishment. Add to these concerns Saudi Arabia's pivot to Russia and economic ties with China.

As I argue elsewhere, "Another point merits emphasis: The political debt that Iran owes Russia and China cannot be paid only in economic matters (energy contracts) but also in strategic vision. The possibility exists of Iran entering into a military treaty of some sort, clearing the way for China and Russia to establish military and naval bases on Iranian soil, similar to arrangements the United States has with Bahrain."[25] This rapprochement between Russia and Iran will be tested in a host of areas, including the civil war in Syria, the turmoil in Yemen, and the targeting of ISIS. While I cannot predict the future of relations among nations, I believe Russia will remain flexible in its strategy in the Middle East, a region changing by the hour, where ambiguity and chaos are the orders of the day.

Iran, on the other hand, sees Russia as a bulwark, especially with its veto power at the UN Security Council. Iran's strategy proved successful when Russia, on many occasions, voted against implementing added economic sanctions on Iran. Further, Iran understood early on the role Russia plays in international forums, including in Iran's recent nuclear negotiations. Hence, it was expected that Russia would back the West as long as its interests were taken into consideration. Iran's assumption was on target.

Iran and Russia forged their cooperative relationship even before the nuclear negotiations. To circumvent economic sanctions, Iran benefited from selling its crude oil to Russia. This economic undertaking provided the Islamic republic with the opportunity to use the proceeds to purchase and upgrade its military hardware. That is no more evident than when Russia delivered to Iran the Antey-2500, an advanced defense system, instead of the S-300.

Does this mean Russia and Iran are about to embark on a military treaty now that the nuclear negotiations are over? There is a strong possibility. As I argue elsewhere, "Russia . . . unexpectedly shelved delivery of its S-300 air defense system to Iran and replaced it with one that is more sophisticated, more powerful. The Antey-2500 defense system was developed in the 1980s and can engage missiles traveling at 4,500 meters per second, with a range of 1,500 miles. This strengthens Iran's position in nuclear talks with the United States. A recent visit by Russian defense minister Sergei Shoigu to Iran even suggests the possibility of military cooperation between the two countries."[26]

As stated earlier, both Russia and Iran realize that, despite their cooperation, their objectives differ significantly. This is especially evident in Syria. While Russia supports the Syrian regime by providing military weapons and economic incentives, Russia's objective is geopolitical. However, Iran's support for Syria is not only economic but also ideological. Further, Iran uses Syria as a platform from which to communicate its foreign-policy vision. One can soon assess whether Iran will maintain this strategy or use its own platform given the possibility of reintegrating into the international community following its historical nuclear agreement with the West. However, the question international security analysts, and citizens in the West, for that matter, are asking is, "Will the nuclear deal bring Iran out of its isolation and allow it to integrate into the international community? Or will Iran continue to be a pariah?" We are not in a position to answer these questions at this time. That said, Iran's support for the al-Assad regime, and nonstate actors including Hezbollah in Lebanon and Houthis in Yemen, will continue whether or not it reaches an agreement over its nuclear program.

Middle East Regional Powers and Growing Tensions

Given the ongoing turmoil in the Middle East and the much-anticipated shift in the geopolitical landscape of the region following Iran's nuclear agreement with the West, other state actors will indirectly influence the outcome of this shift. Among them are Turkey and Saudi Arabia, Sunni countries. The next section addresses each country separately while providing insights into why Turkey and Saudi Arabia stand to augment the debate over what the new political landscape in the Middle East may look like.

Turkey, a secular Sunni Muslim country, finds itself at the center of the ongoing civil war in Syria and the Western coalition air campaign against ISIS. While, earlier in the campaign, Turkey expressed its reservations about joining the fight against ISIS, it reversed its decision in the aftermath of a terrorist attack in the country's southeast. Turkey's participation in the air campaign against ISIS provided an opportunity to target the Kurdistan Workers' Party (PKK), given the long, violent history between the group and the Turkish government; the two-year peace process between the Turkish government and the PKK collapsed in July 2015. Ankara now uses the war in Syria as a pretext to target PKK.

Yet, Ankara claims the reason for going after both ISIS and PKK is to crack down on suspected militants returning from Syria.[27] Turkey's latest decision has complicated matters for the Washington establishment. The White House has now decided to avoid criticizing Ankara in the hope of maintaining Turkey's air support. Further, NATO responded to Turkey's immediate request for a NATO summit. That suggests that NATO members turn a blind eye to the violations Turkey is committing against the PKK. This comes on the heels of the latest statement issued by Iraq's Kurdish regional government calling for the PKK to withdraw from its territories.

The other argument suggests that the flood of refugees coming across the border from Syria has not only triggered a humanitarian crisis but has also created an opportunity for Erdogan's government to expand its air campaign beyond the stated objectives. What impact will Turkey's actions have on geopolitics in the region? The answer lies in Iran's, Russia's, and China's reactions to these developments. Of interest in this discussion is Turkey's downing, in November 24, 2015, of a Russian fighter jet that allegedly violated its airspace. The incident did not result in the outcome many in the West feared; yet, it could be only a matter of time before Russia develops a strategy in response to this incident. Turkey's action prompted some discussions within NATO, behind closed doors, as to whether Turkey is becoming a liability rather than an asset for the Western alliance. As it stands, Russia's military operations—which target Syrian rebels, and, to some extent, ISIS—show no sign of abating. I am convinced that Moscow realizes that it is in its own interests, at this time, not to escalate tensions with Turkey over the downing of its fighter jet. Doing so allows Russia to focus on its main objective: preserving al-Assad's regime.

Yet, the Western coalition objective is the defeat of ISIS, which differs from that of Turkey, which is the complete riddance of the PKK once and for all. Another of Turkey's clear objectives is the removal of the al-Assad regime. However, this aspiration proved to be too challenging given Iran's and Russia's vehement opposition to Western intervention. While the West wishes to remove al-Assad from power through means other than military intervention—which is out of the question—Turkey expresses its vehement opposition to the West's support of Kurdish fights, for instance.

To the rest of the world, Turkey appears to fight a common enemy, ISIS; however, to those with a much deeper understanding of the geopolitics of the Middle East, Turkey is hitting two birds (ISIS and PKK) with one stone (air strikes).

While Turkey claims to want to establish an ISIS-free zone in northern Syria, its objectives extend beyond mere statements and declarations from the Turkish ministry of foreign affairs. Stated differently, Turkey's overall goal is to expedite the process for removing or overthrowing the al-Assad regime, an objective it shares with the Kingdom of Saudi Arabia. While Turkey and Saudi Arabia share similar interests regarding the regime change in Damascus, they have differing objectives. For instance, Turkey wants to promote the Free Syrian Army (FSA), based in Turkey, a group it has been supporting for years. This way Ankara ensures that, when the conflict is resolved—which I do not foresee without factoring al-Assad into the equation—it could convince Washington to support and arm the FSA. Ultimately, the long-term objective is religious in nature: Turkey desires to see the extension of the Sunni Muslim Brotherhood in Syria.[28] Could this explain the sudden about-face in Turkey's policy in joining the coalition against ISIS?

In the following sections, I address in great depth how Turkey's security interests in the region intertwine with those, for example, of Saudi Arabia, not only from a geopolitical perspective but also from a religious angle. Further, as the civil war drags on and the fight against ISIS shows no sign of abating, the challenge for the United States is to decide whether Turkey's support hinders Washington politically in the long term. If so, the US foreign-policy establishment will be forced to go back to the drawing board to figure out what strategy, if any, it should pursue given the unexpected shift in Turkey's position.

The upcoming US presidential election also merits attention. The world, including the Middle East, waits to see whether the next president of the United States will be a Republican or a Democrat. The outcome raises one of two important questions. If the next president is a Republican, does this suggest the United States will embark on yet another senseless, long, costly, and bloody conflict in the region, especially if the US Congress rejects the Iran nuclear agreement? If the next president is a Democrat, does it mean that ambiguous US foreign policy in the region when it comes to ISIS, civil war in Syria, sectarian violence in Iraq, double-standards approach to Egypt, and upheavals in Yemen will remain unchanged?

Considering the first option, engaging in a military conflict with Iran would certainly expand the conflict beyond the region's borders, assuming a Republican president would most likely embark on such an undertaking. Other major powers, mainly China and Russia, would get involved. European allies would oppose the United States' efforts because of lucrative economic incentives from doing business with Iran, since some European sanctions on Iran have always been independent from those of the United States. Further, China would forbid interruption of its energy sources, mainly in the Middle East. The question is whether China would exhibit a military posture should conflict in the Middle East disrupt its access to energy. It is very challenging to state with a degree of certainty what might occur should this scenario become a reality. After all, China—as much as Russia—can be unpredictable if it needs to be; and, given its economic preeminence, military upgrade, and buildup in the South China Sea, there is no telling what may happen.

The second consideration, if the next US president is a Democrat, is whether his or her foreign policy should be more assertive or maintain the status quo (i.e., armchair delegation from the White House). Whatever the presidential outcome is, conventional wisdom suggests the Middle East will continue to pull in the United States despite the ongoing upheavals and the much-anticipated shift in the political landscape of the region. What's important to understand is whether the US foreign-policy establishment is creative enough to come up with new alternatives to its failed and ill-defined policies since 9/11. The United States' ability to articulate its vision for how it intends to cooperate with other global powers and regional players is paramount. However, many wonder whether the United States can refrain from conducting a double-standard foreign policy—one that supports tyrants and dictators (like in Egypt) while at the same time claiming to support free elections and human rights. The United States' standing on these issues, to a degree, would play a pivotal role in providing it the opportunity to reclaim its global leadership that other countries aspire to emulate in many ways.

Detailing what this strategy entails will allow allies like Turkey to plan accordingly and engage with the United States on multiple fronts. At the same time, Saudi Arabia's role in Middle Eastern politics has one objective, which is to limit the expansion of Iran's sphere of influence in the aftermath of the historical nuclear agreement with the West. In the next section, I present the argument of why the role the desert kingdom plays in Middle Eastern politics will have ripple effects on a host of issues ranging from economics and politics to religion and demographics.

This introduction has taken the topical approach to addressing the challenges arising in the Middle East due to a host of issues, including the ongoing civil war in Syria, the sectarian violence in Iraq, the upheavals in Yemen, the failed state in Libya, and the conquests of ISIS, among others. The United States faces major political challenges in the region as regional powers compete for political power, religious dominance, and ideological influence. Of interest is the rivalry between Saudi Arabia and Iran. This section presents an intellectually challenging framework and logical deliberation of how Saudi Arabia perceives the inevitable shift in the geopolitical landscape of the region. Subsequent pages delve deeper into the root causes that fuel Saudi Arabia's concerns and intensify its anxiety over retaining the upper hand in this Cold War–like rivalry with Iran. This emerging cold war appears unconfined to military dimensions; at this moment, it is in reality wrapped within a religious narrative.

Saudi Arabia's anxiety began before the West and Iran's agreement over the latter's nuclear program; it arose rather in the events of the Arab Spring that saw long-established dictators like Hosni Mubarak removed from power. Thus, the kingdom feared that a similar revolt could sweep its territory, mainly in the kingdom's eastern Shiite-dominant provinces including ash-Sharqiyyah and the city of al-Qatif. This anxiety suggests that Shia-Sunni enmity, which emerged following the death of Prophet Muhammad in 632 AD, indeed divides the Middle East along religious lines. Iran's historic nuclear deal with the West is surely feeding that narrative as Saudi Arabia finds itself dealing with too many conflicts: Yemen, civil war in Syria, decline in oil prices on the international market, and the conquests of ISIS, among others.

While these conflicts contribute to the overall apprehension, the kingdom's real issue is its strong opposition to relinquishing its religious dominance in the Muslim world. And how could it when religion is the only currency that legitimates the kingdom on the world stage? Yet, the international community should understand that Saudi Arabia's concerns over the welfare of the Muslim world—or the notion of Sunni unity, for that matter—is an illusion. Great division exists within Sunni countries in the Muslim world for many reasons: some have economic origins, while others trace their root causes to lack of education, inequality, and lack of basic democratic infrastructure.

Other reasons for this division are strictly political. For instance, Egypt, which witnessed the successful removal of dictator Hosni Mubarak through demonstrations as part of the Arab Spring movement that swept most of the Muslim world, descended into chaos that led to a military coup, in 2013, by General al-Sisi. Replacing one dictator with another continues to hinder both the political and security stability of the country. Libya is another example: following the uprising that saw the ouster of Muammar al-Gaddafi, the country fell into civil war. Now Libya is a failed state where ISIS stages its operations. These examples suggest that religion was not at the heart of the Arab Spring; rather, it was social injustice, unemployment, corruption, and social inequality.

I find it impossible to accept the argument that Saudi Arabia fears Shia dominance of the Muslim world. While the majority of Shiites reside in Iran, a country of roughly eighty million inhabitants, this represents only a fraction, along with other Shiites in other countries, of the far larger Sunni Muslim population. Saudi Arabia's rulers do what Arab rulers typically do to reinforce their firm grip on power—they excite the masses through religious narrative and use religion as a means to achieve their objective of total control. History recalls a time when countries like Libya, Algeria, Egypt, Tunisia, Sudan, Turkey, and others used the Arab-Israeli conflict to mask their economic failure, corruption, and limited democratic progress, among other issues.[29] Alas, these days, the narrative some Muslim countries adopt is more about Iran, ISIS, and so forth.

In the case of Saudi Arabia, the difference lies in whether Iran, armed with a nuclear weapon, would be capable of advancing its political, ideological, and religious agenda more quickly. I personally do not subscribe to this notion. Iran will first want to improve its shattered economy following decades of devastating economic sanctions. Once the economy improves, as one may expect, Iran will take other avenues to expand its influence in the region. The West should not be naïve about Iran's aspirations in the region. The latest historic agreement over Iran's nuclear program will certainly enable far greater cooperation between China, Russia, and Europe, mainly Germany and France. As a result, Iran sees building its economy as the foundation on which, economically, it can compete globally, and politically, it can expand regionally.

Against this backdrop, global-affairs and international security analysts wonder whether the desert kingdom may be seeking other avenues to shore up support for its cold war with Iran. Of interest is the kingdom's effort to convince Sunni countries like Jordan, Turkey, Egypt, Sudan, Algeria, Morocco, and others to form a coalition, a bulwark if you will, to prevent Iran from expanding its Shia ideology within the countries listed above. Some observers argue that it may not be a bad idea for the kingdom since all it has to do is provide financial assistance to these countries to get them on board. Yet, the danger in this proposition is that Saudi Arabia may have to rely on Islamic groups like Hamas to advance its agenda.

The security and political landscape of the Middle East has been unstable. Many factors have contributed to this volatility, including the aftermath of the terrorist attacks on the United States in 2001, the invasion of Iraq, the war in Afghanistan, the ongoing civil war in Syria, the turmoil in Yemen, the failed state in Libya, security and political flux in Egypt, and the conquests of ISIS. These events have indirectly shaped Saudi Arabia's foreign policy and, at the same time, have pushed the kingdom's extreme Wahhabi ideology, a form of extreme Sunni Islam, to the forefront.

These dynamics may make some Sunni countries reluctant to proceed given that internal political forces could influence the outcome of this rapprochement. For instance, let's assume Saudi Arabia decides to fund and support Hamas to undermine Iran. Would this support translate into supporting the Muslim Brotherhood since Hamas is affiliated with the former? Further, how might Egypt respond to these developments given the security and political instability caused by the military coup that ousted the only democratically elected president in the Arab world, Mohamed Morsi, who happens to be a member of the Muslim Brotherhood?

The banning of the Muslim Brotherhood from the political process casts doubts on whether democracy will ever take hold in the Muslim world. The short answer is no because there is no separation between church (mosque) and state (government). Frankly, I do not see how any ruler, whether in Saudi Arabia or any other Arab country, will relinquish his firm grip on power for the sake of democracy. Those who think the invasion of Iraq or the Arab Spring paves the way for democracy in the Middle East and the Muslim world entertain mere fantasies. Against this backdrop, one argues how the West, including the United States, failed to support the principles of democracy when the military coup in Egypt ousted a democratically elected president.

Whether you agree or disagree is beside the point. A major assault has taken place on democratic principles, and the only thing the West did—the United

States included—was offer meaningless bravado and absurd prating about free-dom, democracy, and voting rights. I understand that in international relations context matters, but I also understand that democratic principles matter, and *credibility counts*. The United States has a moral obligation to support and defend those principles it was founded on and for which, throughout its history, it has sustained political turbulence. Even people in the streets of some far-off Middle Eastern country can recognize hypocrisy when it appears.

The bottom line is that Saudi Arabia will (a) do whatever is in its power to limit Iran's influence in the region, and (b) ensure that its religious dominance over the Muslim world continues. Regarding the first assertion, the kingdom could embark on a more assertive foreign policy; for example, its military op-erations in Yemen have thrown one of the poorest countries in the Arab world into a civil war for years to come. Could this undertaking lead to a full-blown military conflict between the kingdom and its archenemy Iran? The answer is that it's unlikely because the consequences would be felt beyond the borders of both countries. As Frida Ghitis writes, "The conflict in Yemen now threatens to bring all those simmering disputes to the surface. Neither Saudi Arabia nor Iran wants to enter into open confrontation. But armed clashes and popular passions always risk taking on a life of their own, escalating beyond the calculated wishes of political leaders."[30]

Addressing the second assertion, the kingdom wants to ensure its religious dominance. It has played a dominant religious role since the eighteenth centu-ry, following the alliance between the founder of the al-Saud dynasty and the Wahhabi movement leader, Muhammad Abd al-Wahhab. This alliance formed the foundation of current Saudi Arabian dynastic rule.[31] Further, the kingdom's religious role represents the only currency that provides it with legitimacy. For this reason, Saudi Arabia has put at the top of its agenda vehement opposition to relinquishing any religious leadership in the Muslim world.[32]

Against this backdrop, I find myself asking the following question: Will Sau-di Arabia pursue nuclear technology to demonstrate its willingness to challenge Iran—and the West, for that matter? Yes. As a matter of fact, not only will the kingdom embark on this venture, so too will Jordan, Egypt, Turkey, and United Arab Emirates, all of whom have expressed interest in pursuing such technology. Didn't Pakistan and India engage in a nuclear arms race that ended with both acquiring the bomb? Could a similar fate await the Middle East? The possibility is there, especially with unlimited access to financial resources to acquire such technology.

The challenge for the West lies in whether the time has come for other coun-tries to acquire nuclear weapons independently of the United States. Stated dif-ferently, when Iran acquires the bomb, will key regional players in the Middle East follow suit? I see no reason countries like Saudi Arabia, Egypt, and Turkey would not pursue a nuclear path that will achieve two objectives: (1) challenge

Iran and (2) tilt the balance of power by changing the geopolitical equation in the Middle East in a volatile way. As Reback Gedalyah writes, "Some say the Iranian nuclear program is not a threat in its own right, but that it could break the NPT. Egypt and Turkey might pursue a weapons program out of prestige; Saudi Arabia out of concern for an existential threat. But there is a chance Saudi Arabia might immediately buy one from Pakistan. But Egypt is not in a financial position to withstand the sanctions for pursuing a bomb."[33]

One concludes, how, for instance, Saudi Arabia has suddenly adopted a policy on Russia: Invitation of President Putin to visit the kingdom in addition to the Saudi Arabia's latest investment of $10 billion in Russia. Whoever thought the kingdom, a strong US ally, would pivot toward Moscow? Whether because of Iran's nuclear agreement with the West or the shift in Middle Eastern geopolitics, the Saudis' new policy shift betrays not only the subtle diplomatic rift between Riyadh and Washington but also the decline of US leadership in the region. Saudi Arabia's engagement with Russia—and China, for that matter—suggests its strategic vision in recognizing the unfolding commercial and financial shift toward Asia, where the economic preeminence of China continues to feed growth and create unlimited opportunities for corporate interests, mainly in the energy sector.

Saudi Arabia's investment in Russia is not limited to the energy sector but includes the agricultural sector, health care, retail industry, transport, and real estate; the diversity of this investment is behind the kingdom's latest announcement that it will invest $10 billion in Russia. While most of the diplomatic exchange between Moscow and Riyadh has thus far focused on an exchange of opinion on strategic issues, mainly the civil war in Syria and ISIS, these negotiations have not delved deeply enough to address issues where the countries differ significantly over Saudi Arabia's ongoing military operations in Yemen, for instance. Russia and Saudi Arabia are at odds over whether to keep the al-Assad regime in power. Russia vehemently opposes any form of regime change in Damascus, while Saudi Arabia argues that there can be no solution to the Syrian crisis as long as al-Assad remains in power.

For the relationship between Moscow and Riyadh to develop, to mature, the two countries must agree on solutions to the issues mentioned above. Could that need explain the high-level visits of Saudi officials to Moscow more recently? The Saudi deputy prime minister and minister of defense, Mohammad Bin Salman al-Saud, visited Moscow in June 2015. A month or so later, the Saudi prime minister, Adel al-Jubeir, held talks with his Russian counterpart, Sergey Lavrov. Rapprochement between Russia and Saudi Arabia not only demonstrates US failure in the Middle East but also suggests that Russia has outplayed the United States in the region.

The basis for my latter assertion is twofold: (1) the failure of the United States in the region is evident as the Pentagon fell flat on its face when, after years of training Syrian rebels and wasting millions of dollars of taxpayers' money, Washington had to release the new Syrian forces that ended up in the hands of al-Nusra Front, an affiliate of al-Qaeda, according to Tereza Spencerová, an analyst at the Czech newspaper *Literarni Noviny*;[34] (2) the speed of diplomatic developments and exchange of high-level visits from both countries demonstrates how Russia convinced the kingdom to reconsider its policies pertaining to ISIS and civil war in Syria. This sudden diplomatic maneuvering speaks volumes about US failure to understand the changing regional dynamics. It also betrays Washington's lack of strategic vision. In my opinion, one may attribute this shortsightedness to ambiguous US foreign policy, a policy that requires more than a handshake with key leaders in the Middle East.

Will the relationship between the desert kingdom and Russia endure the ever-changing dynamics and shifting allegiances and loyalties in a volatile Middle East? The answer is anyone's guess since pundits, scholars, and international security analysts think that Iran's nuclear agreement with the West will inevitably shift the geopolitical landscape of the Middle East. Of interest is whether the kingdom will enter into a treaty that includes military provisions; there is no telling. Analysts should make cautious assessments and not read too much into this unexpected pivot of the desert kingdom toward Russia. As I argue in my previous writings, "No matter how this rapprochement plays out, international security analysts shouldn't see this sudden pivot of the Saudis toward Russia through cold war lenses, where a country had to belong solidly to one bloc or the other."[35]

However, this tumultuous environment blinds US foreign policy in the region. How can the United States act in a consistent manner when it does not even have a clear, objective policy in the first place? How can the United States act decisively and consistently when it does not have a clear, objective policy in place and when it says one thing and does another? How can the United States act at all when it lacks credibility in all areas? No matter how it articulates the narrative about the changing dynamics in the region, the United States' failure to secure a decisive military victory in Iraq and Afghanistan has cast doubt on its leadership and resolve. Could this explain why Saudi Arabia, a strong US ally, has finally decided to shift its attention toward the East, mainly to Russia and China? Yet, one must consider the kingdom's future economic interests in Asia when analyzing this sudden shift.

In the subsequent pages, I detail what the shift in the geopolitical landscape in the region could mean for Saudi-Russia long-term relations, assuming the chaos in the Middle East and the tensions between Moscow and Washington remain. In the meantime, other countries in the region play a pivotal role in affecting the Middle East's geopolitical landscape. Among them is Turkey. In the following pages, I present a framework for how Turkey, a secular Muslim country and a NATO member, tries to strike a balance between its strategic aspirations and security concerns.

Before I delve deeper into this theme, let's first address the geography of Turkey. Turkey's location shows why it plays a pivotal role in international geopolitics and why it continues to draw the strategic interests of the West, East, and even within the Middle East. Turkey borders eight countries: Bulgaria to the north, Greece to the west, Georgia to the northeast, Iraq and Syria to the south, Azerbaijan and Armenia to the east, and Iran. This fact gives it significant geostrategic importance, evidenced by its membership in NATO since 1952 and its status as one of Washington's closest allies, at least in theory.

One must also address the historical events that have shaped this country since the glorious days when the Ottoman Empire ruled large swaths of the Muslim lands from the north and the horn of Africa to Southeast Europe and Western Asia. This empire drew some of its strength and influence from the religion of Islam, which was, at some point in its history, the unifying factor. As Hugh Kennedy, a historian and professor at the Department of the Languages and Cultures of the Near and Middle East at the University of London, argues, "Seventy years after the Prophet's death, this Muslim world stretched from Spain and Morocco right the way to Central Asia and to the southern bits of Pakistan, so a huge empire that was all . . . under the control of a single Muslim leader. . . . And it's this Muslim unity, the extent of Muslim sovereignty that people above all look back to."[36]

However, it was Mustafa Kemal Atatürk, known as the father of the Turks, who called for complete independence following World War I. That act helped him, in 1923, to become the first president of modern-day Turkey. However, his tenure in office was marked by failed economic policies and an authoritarian style that prevented other parties from participating in the political process. This experience left a bitter taste within political circles at the time, which gave rise to a multiparty system including the opposition, the Democratic Party, which won elections in 1950. In this period of its history, Turkey fell into instability, suffering military coups in 1960, 1971, and 1980.

Following the history of conflicts in Turkey, especially in 1974, was its military intervention to prevent Greece from taking over Cyprus. Only ten years later, Turkey experienced one of its most enduring conflicts, one that continues to rattle its political establishment today. The conflict has to do with the activities conducted by a separatist group PKK. This conflict with the PKK continues to shape Ankara's policies in the Middle East today.

Turkey's latest military intervention against ISIS also merits discussion. While the West, including the United States, was under the impression that Turkey's military support is to target ISIS, the reality on the ground suggests that Turkey has different strategic calculations. But why was Turkey initially reluctant to join the coalition against ISIS? The answer is much more complicated than one could imagine. Turkey's rationale for not wanting to join the coalition early on was based on two factors. The first, the political climate in Turkey following the latest election win of the current political party in charge, the Justice and Development Party (AKP—known also in Turkey as Adalet ve Kalkinma Parti), raised concerns among countries in the Middle East. Despite his conviction in 1999 for inciting religious hatred, party founder Recep Tayyip Erdogan pressed ahead and ignored the court's 2002 ruling that he was ineligible to run for office. The main challenge the party faces, however, emerged forcefully when a military coup removed from power the Egyptian democratically elected president, Mohamed Morsi. Ankara condemned the Egyptian military coup. It also highlights PKA's support for the Muslim Brotherhood, a stand that was not well received in Saudi Arabia and the Gulf states. Interestingly, the AKP itself traces its roots to a banned Islamist movement.[37]

The second reason has to do with the reality on the ground concerning the balance of power among key players: Iran, Saudi Arabia, Russia, China, Germany, Britain, the United States, and France. Stated differently, the political landscape of the Middle East after Iran's nuclear agreement will be far different from the status quo the region has endured since World War II. Further, if I may be forthright, the role the United States has played in managing and maintaining a diplomatic equilibrium and a sense of stability and security in the region since World War II will cease to exist. The new political landscape following Iran's nuclear agreement compels major powers to reorient their strategies to ensure economic dominance, diplomatic influence, and military presence.

Iran's nuclear agreement provides this opportunity, and I suspect competition will be fierce among major and regional powers. As previously mentioned, economically, Germany and France have already rushed into visiting Tehran to secure business contracts. Switzerland lifted all sanctions on Iran, and soon full-scale trade between the countries will take effect. Militarily, Russia provided Iran with Antey-2500, a more sophisticated and powerful defense system when compared to S-300V. Diplomatically, Iran is already affecting the political process in the civil war–torn countries of Syria, Iraq, and Yemen.

Turkey's agreement to participate in the coalition mainly targets the PKK's locations, including the ones in northern Iraq. This suggests that Turkey wants to kill two birds with one stone: using air strikes as part of the coalition but targeting ISIS and PKK. I ask whether Washington is naïve to think that Ankara will just offer its unwavering support without first considering its strategic interests. Once again, such thinking demonstrates the United States' ambiguous foreign policy and lack of systematic understanding of the dynamics that continue to shape the geopolitical landscape of the Middle East. Further, regional countries and even those far away, such as Russia and China, have concluded that the US military entanglement in Iraq and Afghanistan has had a tremendous cost in blood and treasure. While these events were taking place, Turkey, for instance, strengthened its ties with Russia. According to John Feffer, "The United States has lost a considerable amount of relative power as other countries have surged economically (China has become the leading trade partner for East Asian countries, [and] Russia is now Turkey's major trade partner)."[38]

One should not be surprised that US policies toward the region have been formulated in a political vacuum marked by inconsistencies and continuation of a foreign-policy double-standard. For now, suffice it to say, Turkey cares less about the United States' needs given the latter's lack of credibility and leadership and inability to dictate to the region what to do.

The new geopolitical landscape in the Middle East is certainly forcing regional countries, including Turkey, to reconsider their alliances, priorities, and strategic objectives. This shift comes at a time when Iran reached an agreement with the West over its nuclear program, an agreement that makes possible major economic transformation as economic competition intensifies given the much-anticipated lifting of sanctions on Iran. In fact, European countries, including France and Germany, have already visited Tehran to discuss their corporate interests (i.e., to establish business relations with Iran). Elsewhere, I note, "Germany has already dispatched to Tehran a large delegation headed by trade minister Rainer Brüderle. France's foreign affairs minister, Laurent Fabius, is taking similar steps. And China is eyeing Iran for major economic ventures in the energy sector."[39] While political squabbling and gridlock in Washington continue to hinder the United States' ability to govern and to project itself as a leader, Europe, Russia, and China strengthen their economic ties with Iran as it moves from isolation into global integration.

Back to Turkey: I wonder where these developments leave Ankara regarding what it needs to do given the inevitable shift in the geopolitical landscape of the Middle East. Will it support the United States' and NATO's strategic vision and interests to counter Russia and China in the region? Will it focus on its own agenda independently of the United States and the alliance? Or will it first pursue nuclear technology given Iran's nuclear accord with the West? I try to answer such questions in the subsequent pages as part of my analysis.

Before concluding this introduction, I argue that Iran represents the center of gravity of this entire geopolitical shift, a shift that will have economic, military, and ideological dimensions that reach far beyond the borders of the Middle East. What one can say about Iran requires more time and space than this introductory section allows. However, I present a framework that includes historical, economic, religious, and ideological trends that have shaped the Persian Empire into what it is today. These trends have played a major role and have become part of the Iranian foreign-policy doctrine, one that allows Iran to play a far greater role than it did in the aftermath of the downfall of the Persian Empire.

Of interest in this section is how the nuclear agreement Iran recently reached with the P5+1 will impact the region. Some effects are positive; others are not. On the positive side, the agreement could prove to be a historic opportunity to bring Iran out of the shadows of isolation and into global integration, as China did in the 1970s. Further, both Iran and other countries will benefit from valuable economic opportunities once economic sanctions are lifted. However, if the agreement turns out to be a historic mistake, it will not only further complicate regional politics for decades to come but also put major and regional powers at odds with each other since each country will pursue its own strategic interests. Russia, China, and Iran will try to undermine the United States. Saudi Arabia and Turkey will want to limit Iran's influence. Syria and Yemen will continue to be hot spots where rivalry, especially between the desert kingdom and Iran, will be most noticeable.

Assuming the agreement is a mistake and the United States Congress votes down the deal or authorizes military strikes, Iran will have no choice but to do what Pakistan, India, and North Korea did—develop nuclear weapons in secret. Then Iran would withdraw from being a signatory of the Nuclear Nonproliferation Treaty, which it signed in 1970, leading to uncertainty, instability, and shifting alliances.

A nuclear Iran in the heart of the Middle East certainly will have a major impact on geopolitics in the region and international relations. The question scholars, foreign policy specialists, global-affairs experts, and defense analysts ask is whether the United States has a policy in place to address the inevitable shift. The other question that keeps coming up within domestic and foreign political circles is whether reaching an agreement with Iran over its nuclear program could lead to a new relationship between Iran and the United States. As Jordan Chandler Hirsch, a visiting scholar at the Columbia Institute for Israel and Jewish Studies, argues, "Philip Gordon, the White House coordinator for the Middle East, told reporters last fall that the United States and Iran 'have the potential to do important business with each other' and that a nuclear deal 'could begin a multigenerational process that could lead to a new relationship between our countries.'"[40] At this moment in time, we may not have answers to the aforementioned questions. In the subsequent pages, I examine the tenets the United States bases its foreign-policy approach on subsequent to the Iran nuclear agreement and how these force-producing dynamics will shape US relations with other major and regional powers.

The other key issue has to do with those countries that have vehemently opposed the West's nuclear agreement with Iran. Chief among them are Saudi Arabia and Israel. While each country has its justification for opposing such a deal, they both agree that a nuclear Iran represents a security threat to the region.

What is interesting about Israel's opposition to this agreement is that Benjamin Netanyahu's resentment of this deal plays out well with hawks in the American Congress who state they will vote down the agreement. Because of this statement, a political showdown between the executive and legislative branches of the US government has ensued, saying more about the United States as a nation than whether Iran can really be trusted. Iran's nuclear agreement has divided the US government at its core, thwarting any opportunity for the United States to exhibit much-needed leadership on the world stage. The rest of the world, mainly countries who supported the deal, will move forward to secure economic opportunities with Iran as they gradually lift their own sanctions.

In the following sections, I address in great depth the rationale behind each country's decision to start reorienting its foreign policy given these force-producing events. As a reminder, the challenges in the Middle East did not occur overnight; the development of American foreign policy in the region in the aftermath of World War II did not happen overnight; and addressing the ongoing upheavals and finding viable solutions will certainly take time, patience, vision, and a well-defined strategy. Make no mistake: the upcoming shift in the geopolitical landscape of the Middle East will be marked by a multipolar system in which the United States plays a less significant role for a host of reasons that range from lack of credibility and a decline in leadership to a foreign-policy double standard and unreliability. These factors compel other major powers, including China and Russia, to challenge US hegemony in the Middle East as more and more key regional players, such as Saudi Arabia, pivot east and consider an alternative to the United States.

2   History of the Persian Empire

It would be a daunting task to cover the five-thousand-year history of the Persian Empire in a few pages. However, one must explain, put in perspective, and illuminate the crucial elements that have shaped modern-day Iran given the events of many millennia ago. It would not do the readers justice if I focus my narrative on only one specific period over another in the history of this extraordinary empire. That said, the need to emphasize the early historical events of the Persian Empire could not be greater. Such a study shows the events of the Persian Empire that shaped, influenced, and elucidated the political, social, and cultural plateau on which many agendas competed.

Although knowing the history of the Persian Empire is vital for a better understanding of modern-day Iran, I do not intend to detail, for instance, the Paleolithic times (ca. 1,000,000 to 12,000 BC) or Middle Neolithic to Middle Chalcolithic (ca. 6,000 to 4000 BC) era, or the Elamites. Rather, my writing focuses on the political and social conditions toward the end of the nineteenth century that made possible the profound transformation within twentieth-century Iran, known to the rest of the world as Persia until the early twentieth century.

Before narrowing my writing on this fascinating topic, it is important to define the territories the Persian Empire covered and place those territories within the context of modern-day Iran. That undertaking helps one to understand the nature of Iran's present foreign policy. This examination also responds to an idea held among global-affairs analysts, academics, and policy makers that Iran is expanding its sphere of influence in the Middle East, an influence that is not only military based but also driven by economics and religion. Without understanding the historical background and foundation on which the Persian Empire expanded its influence, one would find it challenging to understand why today's Iran behaves the way it does.

The name *Iran* derives from the word *Aryan*. The Iranian-speaking populace had gradually moved during the first half of the first millennium BC into a region of the Zagros Mountains. The largest groups there were the Medes and the Persians.[1] The Iranian residents of this era used iron tools and irrigation systems, enabling them to use the land to farm more effectively and successfully from the ninth to the seventh centuries BC.

Cyrus the Great founded the Persian Empire. Cyrus first conquered the Median Empire in 550 BC and then went on to conquer the Lydians and the Babylonians. Under the rule of its later kings, the empire grew, governing vast territories, including Mesopotamia (modern-day Iran, Syria, and Kuwait), Egypt, Israel, and Turkey. Its borders eventually stretched over three thousand miles from east to west, making it the largest empire on Earth at that time. Interesting about this era is that, despite conquering different lands, the Persians allowed those they conquered to maintain their cultural customs and traditions in return for paying taxes. Evidently, the Persians' approach contradicted what other conquerors, such as the Assyrians, did. In addition to cultural tolerance, the Persians extended their tolerance to governmental and religious institutions. On the political front, given the vast territories they occupied, the Persian central government had to assign local rulers called satraps, the equivalent of modern-day governors. The satraps' role consisted of enforcing the king's authority and ensuring the collection of taxes. According to historians, about twenty to thirty satraps governed during the era of the Persian Empire.[2]

On the religious front, the Persians, who followed the teachings of the prophet Zoroaster, allowed those they ruled to worship according to their own religious beliefs and teachings. Since the Persians adopted the religion of Zoroastrianism, one sees why it is practiced today mainly in areas once under the Persian Empire. These areas include small communities in western India, central Iran, and southern Pakistan. Today, adherents of the faith also reside in the United States, Great Britain, and Canada.

While detailing the demise of the Persian Empire goes beyond the scope of this present work, suffice it to say, the downfall of the Persian Empire occurred under none other than Alexander the Great, in 334 BC.[3] Alexander conquered the Persian Empire that stretched from Egypt all the way to the borders of India.

Since this book addresses the political impact a nuclear Iran would have on the geopolitical landscape of the Middle East, it makes sense to fast-forward my narrative to Iran at the turn of the nineteenth century to the present time. What follows is a narrative about the evolution of Iran's political climate, how the past has shaped the present, and the revolutions the country has experienced throughout its history. The outcomes of these revolutions have had a major impact on Iran's politics, economy, and religion, which reflect how modern Iran exercises political will and navigates dangerous waters in a world marked by global chaos and tensions. There are those who claim that Iran is at the center of the global chaos. I argue that Iran's participation in this chaos legitimizes it, giving it its raison d'être and an opportunity to showcase that it is not what many in the West think of as a "backward" society. Rather, Iran wields influence in its own backyard.

One must understand that Iran's revolutions started in order to affect a positive change. However, supporters of this change soon discovered that the revolutions amounted to empty rhetoric and disillusion, resulting in flat-out disappointment. Many Iranians' hopes have vanished into thin air. It is fair to say the populace in Iran has grown accustomed to hearing slogans such as "the Dawn of New Era," or "a Bridge to a Bright Future," or "the Awakening of a Great Civilization." Yet, these revolutions produced social strife, class division, and, above all, ideological tensions between secular entities and religious establishments. Those who pushed forward, carried out the vision, and championed the cause of the revolutions were themselves disappointed with the outcome, leading them to withdraw from politics. That is exactly what happened in 1910. Supporters distanced themselves from politics because they lacked the determination and willpower to stand firm in their beliefs. Indeed, decades passed before authors wrote anything about this epoch and its impact on the political culture of Iran. Ervand Abrahamian, a historian of the Middle East specializing in modern-day Iran, writes, "They did not come round to writing histories of the revolution until the mid-twentieth century. Paradoxically, the relative ease with which the revolution was both made and later unmade was linked to the same phenomenon—the lack of a viable central state. The revolution initially succeeded in large part because the regime lacked the machinery to crush opposition. Similarly, the revolution eventually failed in large part because it lacked the machinery to consolidate power—not to mention to implement reforms."[4]

Similarly, one must understand that, in modern history, Iran experienced *two* major revolutions in the span of one century. Those revolutions underscore the disparity between the rulers and the ruled. The revolutions also reflect on the political dynamics, leading one to wonder about the reasons behind the revolutions. Of interest in this discussion are the two major revolutions: the constitutional revolution of the 1890s and the Islamic revolution of 1979. I address these two revolutions in the subsequent pages.

To understand better how Iran's past shaped its present, one needs to appreciate the role the Qajars played in the Iranian political, cultural, and religious saga. During the Qajars' era, the shah was the dominant figure, exercising an absolute monopoly over all institutions. According to Abrahamian, the Qajars, a Turkic-speaking tribal confederation, conquered modern-day Iran in the 1780–1790s, establishing its capital in Tehran in 1786. The Qajars' governance was similar to the current practice in modern-day Middle Eastern political systems, where the central government in some countries governs through tribal leaders. Instead of depending on a central government (as in modern-day Turkey), or religion (as in Saudi Arabia), or bureaucracy (as in Iraq), the Qajars relied on prominent tribal leaders. Sir John Malcolm, a British diplomat, argues that the shahs appointed tribal chiefs, magistrates, and governors but had to choose well-respected individuals in their own community.[5]

Fast-forward to February 21, 1921. A coup by General Reza Khan changed Iran's political, historical, and cultural trajectory. The era of Reza Khan, who became Reza Shah, ended in 1941 following the Anglo-Soviet invasion. This invasion triggered a chain of reactions, including an alliance between the Pahlavi state and the United States. The latter realized the strategic and economic importance of control over oil fields. During this era, the British government had already established a presence in the greater Middle East; thus, a scenario in which the United States could, for instance, influence oil prices, proved challenging. To preserve its interests, the West, mainly Great Britain and the Soviet Union, entered into an agreement to remove Reza Shah but maintain state institutions. Abrahamian writes, "As Sir Reader Bullard—the British minister who was soon elevated to the rank of ambassador—made clear in his typically blunt and frank reports, the Allies kept his state but engineered his removal in part to curry much-needed favor among Iranians. 'The Persians,' he wrote, 'expect that we should at least save them from the Shah's tyranny as compensation for invading their country.'"[6] A consensus among historians holds that Reza Shah abdicated power to his younger son, crown prince Muhammad Reza. The former was then exiled to Great Britain and later went to South Africa, where he died in 1944.[7] The turnover in power facilitated a new political direction for Iran under the crown prince's leadership. The policy was headed by the United States and Great Britain, and, given the outcome of World War II, it was generally supported in the West.

The Allies decided to maintain the military structure under the command of the new shah (Muhammad Reza Shah) to send a message to the Iranian people that the shah, not the Allies, governed the country. The new shah, however, continued in the footsteps of his father by bypassing his minister of defense and communicating directly with his chiefs of staff and generals in the field, a move that suggested the new shah's mistrust of his top generals. Historians agree that, if not for the Allies, Muhammad Reza Shah would not have survived. These events occurred at a crucial economic time, as oil became the topic du jour within political circles, at least from the West's perspective. Great Britain cautiously took the lead in these efforts. Abrahamian writes, "Bullard, the British representative, politely declined the offer of troops and talked instead of more realistic goals. He wrote that the Allies had agreed to give the young shah a 'trial (period) subject to good behavior, which would include the granting of extensive reforms, the restoration to the nation of the property illegally acquired by his father, and the exclusion of all his brothers from Persia.'"[8]

These force-producing events, including political and economic changes toward the end of the 1940s and the beginning of 1950s, paved the way for the emergence of the national movement headed by Dr. Mohammad Mossadegh. Mossadegh was well known within Iran's political circles, dating back to the constitutional revolution of 1906. As one who served in different positions, including parliamentary deputy to the cabinet minister, Mossadegh had an in-depth understanding of the political climate, both domestic and international. He is best remembered for his ability to champion two main issues: strict constitutionalism in domestic affairs and a strong policy abroad known as "negative equilibrium," which ensured Iran remained independent of foreign influence and domination. The latter policy particularly troubled the Allies since the West's strategic calculations went against Mossadegh's vision and policy. To build momentum for his political aspirations, Mossadegh made his case that traditionalists were dragging the country into political darkness while surrendering to world powers.

Mossadegh's narrative struck a chord with the Allies, who sensed political defiance in his speeches. Of interest was his denunciation of two major propositions: the 1919 Anglo-Iranian Agreement and the 1945–1946 oil agreement with the United States and the Soviets. The former consisted of drilling rights of the Anglo-Persian Oil Company, an agreement never ratified by the religious establishment, the Majlis. It was British Foreign Secretary Earl Curzon who issued the document to the Persian government in August 1919.[9] The 1945–1946 oil agreement dealt with complicated dynamics involving multiple players: Russia, Iran, Great Britain, and the United States. Oil was at the heart of the matter. It all started in 1942 when Iran allowed British and Soviet forces into the country to protect Iran's energy sources, mainly oil, from potential German attack. Under that agreement, the presence of foreign forces was legitimate. However, the United States felt left out; thus, the United States pressed for its troops to be in Iran as well.

The 1942 treaty changed the dynamics by giving all foreign troops six months to leave Iran after the war ended.[10] However, Iran failed to anticipate the tactic of the United States and Great Britain, in 1944, to press the Iranian government for oil concessions. The Soviet Union made clear that it would interpret any such concessions from Iran as a betrayal if they, the Soviets, did not get concessions of their own. In 1945, as World War II officially ended, an attitude of mistrust, ideological confrontation, and undermining alliances resulted between the Soviet Union and the United States as each embarked on its own path. This marked the beginning of the Cold War, which defined relations between the nations for almost half a century. The major powers' behind-closed-doors deals regarding Iran's oil resources prompted Mossadegh to call for the nationalization of the oil industry, a move the Allies perceived as setting a dangerous precedent.

It is imperative to shed light on Mossadegh's academic, political, and social background. Mossadegh got his legal education in Switzerland. This experience provided a solid legal knowledge for promoting Western jurisprudence in Iranian society. He vehemently opposed the establishment of the Pahlavi, which landed him briefly in prison. His imprisonment marked a new beginning in Iranian politics as Mossadegh continued to question the legitimacy of treaties and agreements Iran made with the Allies. For instance, he questioned the legality of the 1949 Constituent Assembly on the grounds the elections were rigged.[11] His stand on various issues earned him the approval and endorsement of the masses, mainly the Iranian middle class. His aristocratic background did not deter him from being well received and perceived as "incorruptible," since he lived his life according to middle-class standards, not his upper-class background. It was only a matter of time before Mossadegh launched his political endeavor by

(a) establishing a political party, the National Front, and (b) rallying political parties and associations within the Middle East. Among these parties were the National Party, the Tehran Associations of Bazaar Trade and Craft Guilds, and the Iran Party. Each party was known for a particular feature. For instance, most members of the Toilers Party were prominent intellectuals.

While Mossadegh's political credentials helped to establish him as a major political figure, he also received religious support, which proved crucial to his success. Of interest is the support of the famous cleric during this era, Ayatollah Sayyed Abul-Qassem Kashani. Why is this endorsement important? The answer lies in the cleric's role during his early days in Al-Najaf, Iraq, where he fought the British as part of the Shia revolt. Interesting about Kashani's influence were his ties to a secretive group within Iran known as Self-Sacrificers of Islam—one of the first hard-line organizations in the Muslim world. Whether Kashani's involvement stemmed from his desire to push the religious agenda and give Iran a religious identity or the fear the country might fall into the sphere of the Soviets' communist ideology is subject to further debate. Ray Takeyh, senior fellow for Middle Eastern studies, writes, "Watching Iran's economy collapse and fearing, like Washington, that the crisis could lead to a communist takeover, religious leaders such as Ayatollah Abul-Qasim Kashani began to subtly shift their allegiances. (Since the Islamic Revolution of 1979, Iran's theocratic rulers have attempted to obscure the inconvenient fact that, at a critical juncture, the mullahs sided with the shah.)"[12]

Ayatollah Abul-Qasim Kashani's involvement has always been shrouded in mystery, specifically the question of whether he and his supporters coordinated their efforts with US and British intelligence agencies. For instance, when the religious establishment, including Kashani, decided to get involved in the political turmoil Iran was enduring, they called on and mobilized their supporters in the streets. The argument continues to be that the West did not orchestrate those ongoing demonstrations as part of a plan to overthrow Mossadegh.

Mossadegh, by contrast, saw the frustrations in the Iranian streets. With the support of the middle class, he organized the masses through petitions and demonstrations calling for the nationalization of the oil industry. His quest succeeded in convincing the Majlis, through societal pressure brought by Mossadegh and his allies, to accept a proposal to nationalize the oil industry. This move placed Mossadegh under the microscope of rivals, who saw him as a threat. The shah, Muhammad Reza, perceived Mossadegh as both a challenge and a menace. The national conversation in the Iranian streets, however, started to shift toward where Iran was headed, suspicion of the British presence in the country, and the Allies' strategic calculations.

Mossadegh provoked a dialogue that allowed Iranians to ask hard questions about their leadership as exemplified in the reign of Muhammad Reza. Those dynamics prompted Mossadegh to act, as prime minister, to appoint key figures to key positions in the hope of pushing his agenda and its programs. He created the National Iranian Oil Company (NIOC), one of his main endeavors. NIOC entered into direct negotiations with the British Anglo-Iranian Oil Company (AIOC) to ease the transfer of control of the oil industry from the latter to the former. With the breakdown of these negotiations, Mossadegh ordered the takeover of AIOC and all its resources—oil wells, pipelines, refineries, and so forth. In retaliation, the British government ordered a blockade of exports of oil from Iran and the freezing of Iranian assets. However, tensions escalated even more when the British government decided to reinforce its naval presence in the Persian Gulf. This step prompted major concerns within the British government: "In a post-mortem on the whole crisis, the foreign office admitted that Mossadeq has been able to mobilize the discontented against the upper class closely identified with the British."[13]

The crisis reached a higher level in 1952, when Mossadegh attempted to change electoral law to weaken the shah and the monarchy. Mossadegh argued that, given his rank of prime minister, he had the constitutional authority to nominate and appoint cabinet ministers, including a defense minister (a term used later instead of "war minister"). Mossadegh's challenge represented a direct threat to royal control of the military, a tradition established in the days of Reza Shah, the father. Mossadegh did not stop there. He embarked on a series of initiatives including purchasing defensive weapons, transferring military officers from one unit to another law enforcement entity, trimming the military budget by 15 percent, and transferring royal assets back to the state. His call to replace the monarchy with a democratic republic convinced the Allies of the necessity of political intervention and sabotage. The Allies' scheme put Iran and the West on a collision course. Historians agree that the outcome of that intervention, a coup orchestrated by the Central Intelligence Agency in 1953, set the stage for why and how Iran acts the way it does today. I address this period in Iran's history in great depth in the next chapter.

For now, suffice it to say the United States' and Great Britain's intervention in Iran was perceived, or at least politically promoted, as saving Iran from falling into the communist sphere. The facts suggest otherwise. Oil was at the heart of the matter. The United States and Great Britain wanted to prevent an Iranian monopoly over energy markets in the Middle East while, at the same time, preserving the oil cartel. As a result, in both Washington and London, the vision was to ensure full control of means of energy production and the sale of oil on the world market. Further, the United States and Great Britain wanted to guarantee access to supply sources of Iranian oil to maintain their economic machines for decades to come. The other objective was to ensure that Iran would not develop, manage, and control its own untapped oil fields and thereby set its own oil prices on the international market.

In a sense, the joint venture between the US Central Intelligence Agency and British MI6 served both countries' long-term strategic interests. Western concerns headed by the United States and Great Britain expanded beyond the borders of Iran or the Middle East. The rationale was that if Iran nationalized *its* oil industry, other countries could follow suit. Of interest are Iraq, Venezuela, and Indonesia, countries that are now members of the Organization of the Petroleum Exporting Countries (OPEC) cartel. Could the West's fear of allowing Middle Eastern countries to control the means of production jeopardize the West's strategic interests, mainly industrialized countries? The answer is yes. For this reason, besides others, the Middle East remains a center of gravity where interests among regional and world powers converge and sometimes collide.

History shows that the 1973 oil embargo crippled Western economies and served as a turning point in the relationship between the West, mainly the United States and Britain, and the Middle East. It also transformed the region into a political ring where geopolitical matches took place. The fighting continues as each round brings more energies and new strategies.

Despite the frustration, historians agree that Britain preferred a strategy of engaging Iran rather than becoming entangled in hostilities. This is no more evident than when the British foreign-policy establishment accommodated Iran by paying it royalties in return for maintaining firm British control over oil production. A lack of consensus between the two countries prompted Great Britain to publicly describe Mossadegh as "eccentric," "erratic," "demagogic," and "volatile and unstable," among other negative descriptors. These statements suggest that Washington and London had a mutual interest in ensuring that control over the means of oil production in Iran remained under the influence of the West. Things came to a head in the summer of 1952 when the conclusion was reached that, as long as Mossadegh was still in power, there would be no compromise. That conclusion left policy makers in both Britain and the United States with the option of overthrowing him: "Only a coup d'état could save the situation,"[14] argued the US ambassador.

Once Washington and London reached a consensus, the CIA and MI6 started planning a strategy for a military coup to oust Mossadegh. Britain's long presence in the region allowed it to develop a network of connections inside Iran, including religious figures, wealthy merchants, former politicians, tribal sheikhs, and former military officers. Those assets helped to carry out MI6's and the CIA's plans. As the events of the coup unfolded, the United States made arrangements with the shah, promising financial reward and protection of his monarchy.

Following this tumultuous era was the "White Revolution," a period that stretched from 1943 to 1975. During that time, following the ouster of Mossadegh, Muhammad Reza Shah tried to rebuild Iran on three basic foundations: the military, the bureaucracy, and the court patronage system. Historians agree that Muhammad Reza at the helm of power was nothing but an extension of his father's rule. One can categorize the difference in terms of events on the global stage at that time, politically speaking. Interestingly, father and son both ruled during eras marked by major political changes on the global stage. For instance, Muhammad Reza, the father, ruled during a fascist era and the son ruled during the height of the Cold War between the United States and former Soviet Union.

Rising oil prices in the global market facilitated Muhammad Reza Shah's ability to rule and contributed to the building of a massive state structure in support of his political apparatus and ambitions. Thus, the increase in oil production in Iran paved the way for the country to become the fourth-largest oil producer in the world. One sees why the West perceived the ouster of Mossadegh as a necessity. Without detailing what the new production level of oil allowed Iran to achieve, financially and economically, suffice it to say that Muhammad Reza Pahlavi generously addressed the military establishment's concerns and needs. He provided substantial salaries, unlimited opportunities for both domestic and international training, upscale housing, and plentiful pensions. The shah wanted to purchase the military's loyalty through incentives to avoid any attempt of a military coup to oust him from power.

Similarly, the shah implemented changes in education. During his rule, the number of educational institutions grew substantially. He also addressed some social issues, such as women's rights to vote and run for office; he also increased the marriage age for women to fifteen. While people in some circles, both domestic and global, saw these as positive changes, they produced new dynamics within Iranian society. A higher social class emerged, one linked in one way or another to the shah's family, higher officials, government officials, and senior military officers.

While the upper social class distinguished itself through influence, power, and financial and economic independence, another social class, the middle layer, was emerging. It included merchants and college-educated professionals. The last category consisted of the employees, salaried middle-class citizens (teachers, administrators, managers, and so forth). These social and economic changes in Iranian society created unparalleled tensions, leading many to speculate about the possibility of a social breakdown leading to chaos and disorder. The gap between the haves and have-nots became more evident, mainly in Tehran. These dynamics led to political tensions not only within the government but also among other social classes. The outcome of these events paved the way for the emergence of two prominent figures: Ali Shariati, a French-educated social scientist, and Ayatollah Ruhollah Khomeini.[15]

The revolution of 1979 followed, taking Iran on a different trajectory, one marked by foreign-policy inconsistencies, sponsorship of terrorists, and, as of late, the agreement with the West over its nuclear program. Many security analysts, academics, and foreign-policy experts argue that Iran's nuclear program will affect global geopolitics and relations among global and regional powers. I go further, stating that the agreement allows Iran to play a far greater role in the Middle East. Once it joins the nuclear club—and it will—there will be no limit to how far Iran will expand its sphere of influence, at least in the Middle East; it may defy the assumptions that marked the history of that turbulent region.

The nuclear agreement Iran struck with the West could spell the end of the ruling hard-liners and mullahs in Iran. The lifting of sanctions will eventually free Iran, mainly its youth, to join and contribute to the international community. Tensions have already arisen between young and old generations, moderates and conservatives, the establishment and Main Street, and, above all, those who want to see Iran get out of the cycle of global scrutiny and sanctions and those who want to keep Iran isolated from the rest of the world. It will get worse for Iran before it gets better. However, the looming question is what will become of Iran when the elites, Islamists, and hard-liners at the helm of power decide to go nuclear.

In the following chapter, I address how the history of the Persian Empire shaped and molded modern-day Iran into what it is today. Can Iran's foreign-policy behavior be traced back to its roots as one of the most influential empires in the history of mankind? Possibly.

One must understand that the history of modern-day Iran involves many aspects, ranging from religious doctrines and social customs to political ideology and economic aspirations. These elements, detailed in the following chapter, have defined, contributed to, and shaped the Persian Empire into modern-day Iran. Understanding these basic tenets help us place Iran within the context of the greater Middle East.

Make no mistake—despite its tumultuous history and conflicting ideologies, Iran remains a political force that shapes the economic, religious, ideological, and political narrative well beyond the confines of the Middle East. Thus, Iran continues to attract world powers, with a magnetic force, to compete over its vast energy sources and economic opportunities. The recent lifting of sanctions, part of Iran's agreement with the West over the former's nuclear program, testifies to this fact.

3   Emergence of Modern-Day Iran

ONE CAN BRIEFLY illuminate how Iran evolved throughout the centuries from its glorious days as the Persian Empire to a modern-day theocratic force in the Middle East, yet, how can one talk about Iran without a basic understanding of its centuries-spanning history? After all, Iran's rich history captures the imagination of intellectuals, fascinates artists, provokes the curiosity of politicians, and even lures conquerors. Iran's complex history involves the existence of different religions within its own borders, social and economic influences, and transformative internal dynamics. Those influences, those changes, are what made Iran what it is today. This is no small change to witness in the social fabric of Iran society, which wore its storied history through decades of glory and triumph and now seems clothed, to some, in the costume of a pariah. A different Iran appears today, one that demands attention *precisely because* of how people perceive it. Questions arise: Is Iran an aggressive power or a victim? Is Iran traditionally expansionist or traditionally passive and defensive? Is the Shiism of Iran quietist or violent and revolutionary? Only the study of history can answer these questions.[1]

First, it has been poorly explained to people in the West that Persia and Iran are the same country. Yet, the history of Persia, known for its romance, elegant and well-manicured gardens, and colorful handmade double-knotted carpets, completely differs from modern-day Iran. In the cliché of the Western media's portrayal of Iran, the name conjures a dark image—frowning mullahs with foot-long beards, women covered head to toe (which, by the way, is non-Islamic, since there is no verse in the Quran that suggests a woman dress or cover herself from head to toe), and angry images of crowds burning American flags.[2]

Iran's distant and recent history has been the subject of many academic books. Of interest in this discussion is the immense seven-volume *Cambridge History of Iran*, which covers the history of Iran in its entirety. My study shall not compete with such a multivolume work. This book seeks to present a well-balanced narrative supported by facts and documentation for those with little or no prior knowledge about this fascinating country.

The narrative herein provides a better understanding of what forces shaped Iran to make it what it is today. Understanding basic historical facts about Iran allows us to place it within the context of the greater Middle East, a region that has always been the center of gravity of the ultimate dreams of conquerors and peacemakers alike. Given that pull, the history of modern-day Iran testifies to

the challenges Western powers endured while trying to control it at the turn of the twentieth century. Those challenges render Iran politically and socially un-predictable, thus introducing many security challenges. Iran also continues to attract world powers with a magnetic force, so to speak, into competition for its vast energy resources and economic prospects. The recent lifting of sanctions, part of Iran's agreement with the West over the former's nuclear program, testi-fies to this fact.

What follows is a brief account of key characteristics of Iran, including its geography, history, religion, demographics, and foreign relations. These charac-teristics show how, throughout the centuries, modern-day Iran has become what it is today. However, readers would be treated unjustly if I were to omit, no matter how brief, a narrative of the historical identity of Iran. Therefore, this chapter begins with that discussion.

Throughout much of its history, Iran was known as Persia. Historians agree that Iran was one of the ancient influential empires of human history. Historical happenings, including the overthrow of the monarchy in the 1970s, helped in-troduce modern-day Iran under the rule of its religious establishment. The 1979 Islamic revolution, led by Ayatollah Ruhollah Khomeini, officially sealed Iran's fate under a religious establishment. Yet, to place this change within the context of the political landscape of Iran, one must address key areas that I believe have contributed to modern-day Iran's transformation.

Geography of Iran

Iran, officially known as the Islamic Republic of Iran, is located between the southwestern part of Asia and the Middle East. To the north, Iran borders the Caspian Sea and Turkmenistan. To its northwest, it borders Turkey, Azerbaijan, and Armenia. On the west, it borders Iraq, and on its long southwestern borders, Iran meets the Strait of Hormuz, which in turn meets the Gulf of Oman. On the east, Iran borders Afghanistan and to the southeast, Pakistan.[3] This geography makes Iran a strategic location that bridges different cultures, allowing it to play a vital economic, political, religious, and cultural role.

Its topography varies as much as its demographics, languages, political views, and religions. Iran is rich in natural resources, including oil and natural gas, and its rugged terrain includes mountains, valleys, and deserts. These geographic features challenge the central government in Tehran to develop those areas to keep up with the economic developments taking place in other parts of the world. Two of Iran's main mountains, the Zagros and the Alborz, stretch long distances of about 1,600 kilometers (1,000 miles) and 600 kilometers (400 miles), respectively.[4]

Some of Iran's most important cities are Tehran, Mashhad, Esfahan, Shiraz, and Qom, among others. Qom holds religious meaning for Shiites similar to the Vatican for Catholics and Jerusalem for Jews, Christians, and Muslims. Most Shiites converge on Qom city where the shrine of Fatima Ma'suma, sister of Persian imam Ali Ibn Musa Rida, is located. Besides serving as a major destination for pilgrims, Qom is considered the capital of Shia scholarship. Qom, which is located about 125 kilometers from Tehran, holds religious significance; it served as a stage from which the Basij—Iran's paramilitary force—launched its operation to crush demonstrators in nearby city of Tehran during Iran's 2009 election. The force managed to maintain order in the urban center, mainly in Tehran.[5] A decade earlier, questions were raised about whether the changes made by former president Khatami, a reformist, would bear fruit.

Iranian youth, mainly students, faced not only public apathy but also government wrath. Hooman Majd writes, "In one of the biggest challenges to the government, at a time when Khatami had ushered in reforms unthinkable to hard-line conservatives, students protest in 1999 led to street riots, causing a level of unrest that conservatives viewed as threatening to the regime."[6] These events underscore the ongoing tensions between hard-liners and moderates. Despite the slow progress, student dissent is certainly gaining momentum. Unfortunately, due to its control over the means of production and the security apparatus, the government has the upper hand.

There are various opinions on the exact date that Iran gained its independence, and most scholars and governmental records offer no "official" date of independence. Iran's independence came under different dates and periods. For instance, Ali Alfoneh, a senior fellow at the Foundation for Defense of Democracies, argues that the Islamic Republic of Iran proclaimed its independence on April 1, 1979. He also notes January 16, 1979, when Shah Reza Pahlavi fled Iran to escape popular political revolt against his rule. One could consider that earlier date the official day of Iran's independence.[7] Other accounts, however, conclude that Iran's independence day goes way back to April 1, 1979 (Islamic Republic of Iran proclaimed); notable earlier dates include ca. 550 BC (Achaemenid–Persian Empire established), 1501 AD (Iran reunified under the Safavid dynasty), and December 12, 1925 (modern Iran established under the Pahlavi dynasty).[8]

Cultural Heritage and Landmarks

No one can dispute that, throughout the history of the world, Iran has helped to advance civilization in areas ranging from art and science to medical discoveries and agriculture. Actually, Iran is home to fifteen United Nations Educational, Scientific, and Cultural Organization (UNESCO) world heritage sites. Other cities, such as Shiraz, Esfahan, and Tehran, represent cultural centers that reflect the rich history of the Persian Empire. For instance, Shiraz is known as the "city of roses" and the "city of poets." Not far from the city, only fifty kilometers (thirty miles), is the ancient city of Persepolis, also known as the ceremonial capital of the Achaemenid Empire, once one of the richest cities on Earth. Archeologists and historians believe that Persepolis was established during the peak of the first civilization in that part of the world.[9]

Another city that reflects historical and cultural richness in modern-day Iran is Esfahan. Given its cultural and religious influence, the city reflects Islamic architecture. The city's main square, known as Imam's Square, was built during the Safavid era (approximately 1501 to 1722). Tehran, on the other hand, is Iran's capital, home to many historical sites. The city of about nine million inhabitants is considered the largest city in Iran. Tehran competes with other metropolitan cities in other regions. For instance, Tehran is the second-largest city in Western Asia and the third largest in the greater Middle East. According to the city's mayor's statistics, Tehran is ranked twenty-ninth in the world by the population of its metropolitan area.[10]

Religious Diversity within the Islamic Republic of Iran

While religion plays a major role in shaping Iran's political system, one need only *look* to see how theocratic Iran's government system is. Iran's religious history traces its roots to Zoroastrianism. Zoroastrianism was the dominant religion in Iran centuries ago, mainly through the eras of the Median, Achaemenid, and

Sassanid empires. However, the religious change in Iran from Zoroastrianism to Islam, mainly Sunni, is attributed to the Muslim conquest of Persia.

Modern-day Iran is considered a Shia Muslim country with 90 to 95 percent officially adhering to the teachings of Islam. While, theoretically, there are no major differences in religious principles and practices between Sunnis and Shiites, the two groups have observable practical and ritual differences. The main issue of difference between the two sects of Islam has to do with the assassination of Husayn, the Prophet Muhammad's grandson, in the seventh century. It is imperative that I offer a historical synopsis that puts the schism between Sunnis and Shiites in its historical context.

It all started when Prophet Muhammad passed away in 632 AD and his followers picked Abu Bakr as his successor and first caliph of the Islamic *ummah* (translated as "community"). The selection of Abu Bakr came over the objection of those who favored Ali Ibn Abi Talib, Muhammad's cousin and son-in-law. At that juncture, the debate over who should rightfully succeed Prophet Muhammad led to division and, ultimately, the emergence of the two main sects in Islam: Shiites and Sunnis. "Shias, a term that stems from *shi'atu Ali*, Arabic for 'partisans of Ali,' believe that Ali and his descendants are part of a divine order. Sunnis, meaning followers of the *sunna*, or 'way' in Arabic, of prophet Mohammed, argued and vehemently opposed political succession based on the prophet Mohammed's bloodline."[11] Ali became a caliph in 656 AD; however, his ruling over the Islamic community lasted only five years before he was assassinated.

A small percentage, 4 to 8 percent, of Sunni Muslims lives in Iran today, including the Kurds and Balochs (Baluchistan). Less widely known is that non-Muslim minorities, including Christians, Jews, and Yazidis, among others, live and worship in Iran. Speaking of Iranian Jews, evidence in the literature suggests the extraordinary cultural contributions Iranian Jews have made to Iran's society. However, after the Iranian revolution of 1979, many Iranian Jews decided to leave Iran and resettle in Israel.[12] Also less widely reported in the West is that Christians still reside in Iran; they total about 117,704, according to Jamsheed Choksy, distinguished professor of Central Eurasian Studies at Indiana University in Bloomington.[13] The government of Iran recognizes and preserves the right of other religions, including Christianity, Judaism, Zoroastrianism, and the Sunni branch of Islam, to be represented in the political process and governmental establishments such as the parliament.

I emphasize the religious aspect of modern-day Iran to put this diversity within the context of its growing religious influence. Few countries in the Middle East tolerate, let alone embrace, such religious diversity. Against this backdrop, the schism between Sunnis (primarily in Saudi Arabia) and Shiites (dominant in Iran) now witnessed in the Middle East goes beyond one country's borders. This next part confounds many Americans: while Saudi Arabia calls for unity in the Muslim world, this notion is a myth. The argument that Sunni countries like the

Kingdom of Saudi Arabia propagates—that Iran is going to take over the Muslim world—is baseless and meritless. Saudi Arabia must stop using religion to justify its ends, whatever they may be. Indeed, the issue is not limited to Saudi Arabia but extends to the entire Muslim world. Do you recall when the Arab Spring erupted how, in the West, we argued that, finally, the Muslim world was on its way to embracing democracy? It turned out to be an illusion. As I have outlined in previous writings, there will never be a democracy in the Muslim world because separation of church (mosque, in this case) and state (government) does not exist. Do you expect an Arab leader to give up power so he can be on equal footing with commoners? The answer is a definite no.

I argue that the religious establishment within countries like Jordan, Egypt, Saudi Arabia, and other Sunni countries saw Iran's revolution of 1979 as a threat to both their right to rule and their religious identity. Simply said, the 1979 revolution provided Ayatollah Ruhollah Khomeini a historic opportunity to pursue his long-term strategy, one that foments the idea of ruling the Islamic government through the "guardianship of the jurist" (*velayat-e faqih*).[14] Could this explain why Khomeini was such a controversial religious figure in the Muslim world? Further, the concept of the "guardianship of the jurist" holds that the religion of Islam gives the *faqih*, Arabic for *imam*, custodianship over people. However, this concept in itself is controversial because religious scholars disagree on what this custodianship role should encompass.[15] One school of thought suggests having a limited guardianship of the jurist to nonreligious matters, such as real estate, financial matters, and so forth. Others support the concept of an absolute guardianship of the jurist, suggesting the imam must have full responsibility in all matters, including governance of the country. I argue that the concept itself is controversial not only among Sunnis, who have adopted the approach of separation between political leadership and religious scholarship, but also among Shia religious scholars. Ayatollah Ruhollah Khomeini promoted exactly this concept through a series of lectures he presented during the 1970s before the Islamic revolution. A decade later, the concept of guardianship of the jurist became the basis of the constitution of modern-day Iran. Thus, the term *supreme leader* became synonymous with *guardianship of the jurist*.

Hidden agendas, political or otherwise—from both sides—drive the animosity between Sunnis and Shiites (categorized as Saudi Arabia versus Iran). Undoubtedly, the Iranian revolution of 1979 revived the Shiite ideology that had lain dormant for centuries. Certainly, this rise will empower an Iran that is armed with nuclear capabilities and a robust economy to influence the geopolitical landscape in the Middle East in many ways: "The Iranian revolution in 1979 finally allowed Shiite ideology to re-emerge in a robust way. Iran's strategic vision since has focused on strengthening the ideology through support to Shiite communities in neighboring countries with Shia minorities. The objective always has been and continues to be for Tehran to enhance its influence and expand its religious ideology. These aspirations are backed by a strong military apparatus to project regional power."[16] Now consider this final point. Saudi Arabia's religious rift with Iran reflects its fear of relinquishing religious leadership in the Muslim world given that the kingdom is home to the two holiest sites in Islam: Mecca and Medina. Moreover, how could the kingdom not fear a loss of religious influence when religion is the only currency that gives the kingdom legitimacy on the world stage?

I hope the narrative offered herein serves as a snapshot of Iran's religious and political aspirations. With that snapshot, one may understand Iran's long-term objectives. The information included in this section should afford the US foreign-policy establishment a better understanding of the history of the Middle East as well as its complex religious tensions, social intricacies, and tribal mentality. The above narrative emphasizes that the region is awash in competing political agendas and that it resists change and distrusts foreigners. Such a perspective should help foreign-policy experts formulate an objective policy with positive consequences that go beyond the borders of the Middle East.

As Iran expands its economic, religious, and—soon—military apparatus, the geopolitical landscape in the Middle East compels US policy makers to refrain from empty rhetoric and acknowledge the blood and treasure the United States has invested in the Middle East with little in return.

Government Structure

The current structure of the Iranian government is theocratic, defined by a close relationship between the religious establishment and the government. This political setting was established when the Islamic revolution of 1979 overthrew Shah Reza Pahlavi and created a new constitution in the same year. The new constitution granted broad powers to the president, who was elected by the people for a four-year term. The establishment of the new constitution made possible some radical changes in the Iranian political landscape. For instance, the office of the prime minister was abolished in the late 1980s. However, after returning to Iran in 1979, Ayatollah Ruhollah Khomeini, the spiritual leader and founder of modern-day Iran, assumed full authority that surpassed that of the elected president, thus granting himself the title of supreme leader. This title provides him wide-ranging authority over the military, intelligence services, and the country's strategic decisions. The terms of Iran's negotiations with the West over its nuclear program, which had to be approved by the supreme leader, show the scope of his power. An interesting fact about Iran's political system, at least concerning how the government operates, is how the supreme leader is elected—handpicked, if you will—by what is termed an "assembly of experts." Of note, the appointment of the supreme leader is for life.

Similar to the three branches of the United States' government system, the executive branch in Iran is home to three separate entities: the Assembly of

Experts, the Expediency Council, and the Council of Guardians. Each entity has a specific mission and performs a particular duty within Iran's political apparatus. For instance, members of the Assembly of Experts, elected by popular vote, officially appoint the supreme leader. The Expediency Council's main function is to conduct checks and balances of other branches of the government, mainly the executive, judicial, and legislative, and the Council of Guardians ensures the observance of legal procedures and their agreement with the constitution and certifies that the laws enacted are based on the Islamic law, also known as Sharia law.[17]

Detailing Iran's government structure would make an excellent topic for another book. In this work, however, it is vital to focus the narrative on how the basic structure defines the different phases Iran has gone through since the Islamic revolution of 1979. However, it is important to emphasize the role Khomeini played in forging the perception that a separation between religion and politics would harm Iran and therefore should be rejected. Doing so paved the way for most, if not all, members of the religious establishment, at least within the government, to interpret and implement Islamic law. Yet, there are those who oppose such an approach and see it as an innovation without support in Shia theology. These sorts of interpretations only support my argument that democracy—or its application, for that matter—will never be implemented in the Muslim world because the concept of separation between church (mosque) and state (government) does not exist in the Islamic faith regardless of the sect, Shiite or Sunni. Barbara Ann Rieffer-Flanagan writes, "For Khomeini a just society required that the clergy play a significant political role. However, the notion of an Islamic cleric at the center of political power in a capacity to oversee the political system was not embraced by all Shia clerics. Some viewed the *velayat-e faqih* as a departure from traditional Shia doctrine."[18]

The assertion above highlights the delicate balance sought concerning religion in politics. Without delving deeper into the religious nature of the question, this section focuses mainly on aspects I believe have shaped Iran's politics, economy, religion, and social sphere to make the country what it is today. Iran's postrevolution leadership has been sustained by its Islamist followers. Those followers have supported the revolution with both strong conviction and powerful ideological reasons, maintaining its momentum. As a result, the political revolution in Iran always justify its anti-western slogans as the basis to ensure the commitment and backing of its Islamist supporters. One particular issue that unified radical Islamists of all stripes was anti-Westernization. On his return from exile in France, Ayatollah Ruhollah Khomeini perceived an important need in Iranian society after the era of the shah. The people had to embrace the new ideology and work toward preserving it at any cost. Said differently, the revolution of 1979 was more than just a change in the political landscape; it was about a basic change to the social, political, and cultural fabric of the society and beyond.

With political shrewdness, Khomeini understood that if he could influence the psychology and arouse the emotions of the masses by identifying the true enemy, he would succeed in establishing a moral and religious baseline from which to expand his newly acquired power. The United States' reaction to the ongoing events during this time provided Khomeini the perfect opportunity to identify the enemy: the United States. James Buchan writes, "Khomeini at last had the enemy he needed, and it was not Mohammed Reza (whom he did not name), nor Parliament, nor Prime Minister Mansur, nor Britain, nor the Jews, nor Bahais, but the United States. 'Let the American President know that in the eyes of the Iranian people, he is the most repellent member of the Human race.'"[19] In the case of Iran, this approach, ensuring complete control over society, was not achieved through the application of force, deprivation, or a crackdown on liberties (though some argued it was). Rather, it was through the religion of Islam. The success of the 1979 revolution was sold to the Iranian people under the banner of Islam. In most, if not all, Arab countries, when a government wants to justify its actions or rally the masses to embrace its policies, all it has to do is use Islam both to justify and support its argument. Iran is no different. After the country's experience during the shah era, Iranian people of all stripes were ready for change, a change that allowed them to live according to Islamic teachings. What facilitates the sale of this narrative is that Iran's society disbelieves in, refuses to subscribe to, and even disallows a separation between church (mosque) and state (government). Shalaleh Zabardast writes, "This is obvious that Iran revolution in a case of leadership, ideology[,] and goals of the revolution was unique. It was under the banner of Islam and was the symbol of returning to the model of Islam and was promising simple, puritanical[,] and authentic life for Muslims. Iran's revolution was not just political but social, economic, and cultural one. It brought up new social, cultural orders and new Islamic revolutionary elites. The outburst of revolution at the center of a multi-ethnic country or at the heart of a plural country had international consequences."[20]

What would follow the revolution was anybody's guess. Iran's bloody war with Iraq (1980–1988) cast doubt over the revolution's success. Simultaneously, the United States was concerned about the long-term geopolitical impact of the revolution. It decided to support Iraq, thinking Saddam Hussein would be able to defeat Iran. To the detriment of the Iraqis and to the chagrin of the United States, the war united Iranians of all stripes, and the eight-year conflict compelled major powers to reconsider their geopolitical calculations in the greater Middle East. However, the war with Iraq introduced internal chaos, political struggle, and economic disorder in Iran. The thinking in Iran after the death of Ayatollah Khomeini, in 1989, was that there might be a possibility for political and social change. The hopes of many, mainly in the West, were shattered as Iran pressed on in its theocratic journey by appointing for life Ayatollah Ali Khamenei as the next supreme leader.[21]

Many security analysts, academics, and pundits agree that the Iranian revolution of 1979 changed the security architecture of the Middle East on multiple fronts and reversed the widely held notion that the region would remain an ideological battleground between the East and the West. Whether or not we want to admit it, the Iranian revolution of 1979 reshaped security alliances. It introduced anxiety and worries in Saudi Arabia and other Gulf states about the future of their monarchic regimes. While relations between the United States and Iran remained lukewarm after the revolution, the hostage crisis led both countries to sever their diplomatic relations. Among the scholars who argue in support of the latter assertion is Seyed Hossein Mousavian. He writes,

> In the aftermath of the revolution, Iran did not sever its ties with the United States, while the US also recognized the new revolutionary government. This was a clear indication that the two states intended to open a new chapter in Iran-US relations. However, this phase was short lived. A few months after the victory of the revolution, radical Muslim students stormed the American Embassy in Iran and took 52 Americans hostage. Few could have imagined that the effect of the hostage crisis would be so long lasting. That was the beginning of a new era in the relationship between Iran and the United States, characterized by intense hostility and mistrust.[22]

Iran's revolution of 1979 rewrote both the United States' and the USSR's geopolitical equations and strategic interests in the Middle East. The United States lost an important ally during the Cold War when it lost Iran to the revolution. The balance of power turned upside down. Seats at the geopolitical table were rearranged, and decisions regarding global affairs, mainly in the greater Middle East, were no longer the call of a multipolar system dominated by the United States and the USSR. Three decades later and, as illustrated by current events in the Middle East, Russia reasserts itself in the region while China increases its diplomatic and economic footprints and the two nations team up and gain a new partner in the region, Iran. This partnership will eventually transcend economic trade and become a security alliance.[23]

The West must not forget that, under the banner of Islam, Iran's revolution showed the Arab world that the highest levels of government could adopt an Islamic political system. That question had been left unanswered by the scene in the Arab and Muslim world's political landscape following the demise of the Ottoman Empire. However, the failure of the Arab Spring, the absence of democratic institutions, prevention of political participation, and massive corruption throughout most Arab countries led the masses to ask, Is the adoption of political Islam the solution to ongoing struggles in the Arab-Muslim world? The answer is not that simple, as one might expect. Thus far, the reality on the ground suggests that the masses in the Arab world are looking to religion to play a role in, but not dominate, the government. Egypt in the aftermath of the ouster of Hosni Mubarak provides a clear example of the prevailing sentiment.

Iran's Legal Framework

The legal system in Iran is structured differently in the way the supreme leader appoints the head of Iran's judiciary. The latter does not have to go before a committee for approval. The supreme leader must be careful as to whom he appoints to this critical, sensitive post. The head of Iran's judiciary has the authority to appoint two key legal positions: the head of the supreme court and the chief public prosecutor. Similar to the structure of the American system, which has the appellate and federal courts, Iran's court system consists of (a) public courts, which deal with criminal cases, (b) revolutionary courts, which deal with specific national security matters, and (c) the special clerical courts, which deal with offenses committed by clerics. Of note, the latter court operates independently of the regular judicial system because it answers directly to the supreme leader. Rarely can a verdict be overturned since the court's rulings are final. Only the supreme leader can reverse a verdict, by intervening.

The Assembly of Experts meets once a year for a week. It encompasses eighty-six "virtuous and learned" religious clerics who are elected for an eight-year term. The assembly elects the supreme leader and has the constitutional authority to remove him from power as it sees fit. Of note, since the establishment of the Islamic republic in the aftermath of the revolution, the assembly has never challenged the decisions, past or present, of the supreme leader.

Military Development and Capabilities

Before addressing Iran's military capabilities, one must shed light on its historical military strength. Doing so paints a clear picture of elements that assist Iran in sustaining the strength of its military apparatus. Of interest in this discussion is how, despite the changes in global geopolitics, mainly the end of the Cold War, Iran remained committed to disallowing the political changes on the global landscape to impact its military readiness. While many countries in both the West and the East reorganized their military structure, number of troops, and so forth, Iran did not follow a similar path. The reason is that Iran's continued concern over, and perceived threats from, hostile nations, including the United States and the United Kingdom, convinced Iran's military leadership to maintain military readiness and high troop numbers.[24]

Iran decided to reduce the number of its revolutionary guard from 170,000 troops at the beginning of the 1990s to 125,000 a decade later.[25] I attribute this reduction to Iran's strategic planning, specifically to its realization that, especially after the 9/11 attacks, terrorism issues would occupy the United States. Iran thought it made sense to raise the number of other units, including the army, which increased by about 15 percent. The air force and air-defense forces combined saw their numbers increase by 50 percent toward the early years of the twenty-first century.[26] These changes suggest that Iran's main military concerns had to do with the air rather than the ground. Said differently, Iran feared that the United States, along with its allies, might conduct air strikes (as they did in Afghanistan after the 9/11 attacks) rather than a ground invasion (like the US invasion of Iraq, which immediately followed air strikes).

Western military intelligence estimates put Iran's military budget in the range of $14 billion.[27] There are, however, major differences between the estimates provided by the International Institute for Strategic Studies, an American think tank, and the Stockholm International Peace Research Institute, a Swedish institute. For instance, the former suggests that Iran outspent the United Arab Emirates by about $4.7 billion for a total of $18.1 billion. By contrast, the Stockholm International Peace Research Institute suggests that Iran spent only $12.7 billion, a difference of $6.3 billion when compared with the United Arab Emirates.[28] Whatever the actual expenditure is, international security analysts agree that Iran's military capabilities have become outdated due to the crippling effects of the sanctions. As a result, Iran did not have the strength to compete with its neighbors, considering, for instance, Saudi Arabia's defense spending or that of Kuwait and Egypt among others.

Understanding what was at stake, Iran hesitated to consider military ventures in other countries in the region not only because of its inability to modernize its military apparatus but also from fear of a much stronger force overwhelming it. For instance, Iran could have intervened militarily in Yemen when the Houthis, a Shia rebel group, came under attack. Iran understood early on that it was in its interest not to send ground troops but to provide military, financial, and logistical support to the Houthi rebels. The conflict in itself reflects a regional struggle between Sunnis (led by the Kingdom of Saudi Arabia) and Shiites (led by Iran).

Equally important, Iran realizes that it is in its strategic interest to avoid physical involvement in conflicts. Rather, it makes sense to provide logistical, financial, training, and spiritual guidance to groups like Hezbollah that are willing to carry out missions on Iran's behalf. For instance, the three-year investigation of the Khobar Towers bombing attacks on June 25, 1996, that targeted a US military housing compound in Saudi Arabia revealed that Iran was responsible for providing logistical and material support and manpower to conduct the attacks.[29]

Once Iran realized its ability to support attacks while avoiding any associations, it engaged in more sophisticated operations of prepping agents, including Imad Mughniyiah, who, according to American and Israeli intelligence services, was involved in a series of bombings, kidnappings, and assassinations. Mughniyiah was able to evade capture for almost a quarter century. However, on the night of February 12, 2008, a powerful car bomb detonated as he was passing by on foot in the Kafr Sousa neighborhood of Damascus, Syria. The car bomb ended his life.[30] These events suggest that Iran's support of terrorism activities represents one of its foreign-policy tools.

Yet some events have not been reported accurately or have been altogether misrepresented. I believe Western media, conveniently or ignorantly, refrained from disclosing that Iranian radicals and their Lebanese allies had revealed the secret US-Iranian contact and the kidnapping of three US citizens in January 1987.[31] The Iranian radicals disclosed this information in order to undermine the moderates' efforts to end hostilities between Iran and the United States, which would improve relations with the West.

Why are terrorist tactics one of Iran's foreign-policy tools? The answer depends on whom you ask. Regardless of the answer, under no circumstances must one underestimate Iran's ability to influence events on the ground. As I have argued, Iran's clandestine military activities in Yemen in support of the Houthi rebels highlight the reason the rebel group is able to fight despite the Saudi-led Arab coalition air strikes. Like the Hezbollah fighters, who used to travel to Iran for training, the Houthis are now traveling to Iran for training and financial assistance. According to David Weinberg, a Yemen watcher at the Foundation for the Defense of Democracies, a think tank in Washington, if not for Iran's support, the Houthis would have been unable to conquer Sanaa, the Yemeni capital.[32]

Similarly, the support Iran continues to provide the Syrian regime in the ongoing civil war testifies to Iran's desire to influence the political outcome to its favor. Another illustration that highlights Iran's meddling in the internal affairs of its neighbors is Iraq. This is evidenced by the increase in sectarian violence fueled by the support Iran provides the Shiite militias, especially regarding the situation in Tikrit. This underscores Iran's deep involvement in Iraq, increasing sectarian tensions and security instability there.[33]

While Iran's support of Hezbollah is no secret, it is no more evident than in the Syrian civil war, where members of Iran's revolutionary guard and Hezbollah fighters have coordinated their efforts to attack the rebels' locations. While Russia provides air support for the al-Assad regime, Hezbollah fighters are on the ground, thus introducing a new approach to conventional warfare. Major-General Noam Tibon, commander of the Israeli Defense Force Northern Corps, stated at the Jerusalem Post Diplomatic Conference in Herzliya, "We must understand that this is not a war in Syria where Syrians are fighting against Syrians. Hezbollah . . . is the elite force today fighting against the rebels in Syria."[34] I think the preceding example is a legitimate reason Arab leaders in the region are nervous about Iran's approach to Middle Eastern affairs and its desire to expand its sphere of influence. Because of this, many see Iran's nuclear agreement with the West as a path for Iran to acquire more influence militarily, economically, and ideologically.

Have embargoes imposed by the West undermined the development of Iran's military capabilities? Theoretically, the answer is yes. However, reality suggests otherwise. On the one hand, the embargoes compelled Iran to find ways of modernizing its weapons through domestic productivity. On the other, the embargoes made Iran vulnerable to countries like Russia. Russia sees an opportunity to support Iran's military efforts in return for the provision of a platform on which Russia can forcefully reassert itself in the Middle East. I address the last assertion in great depth in the subsequent chapters. For now, suffice it to say that Iran's missile technology, for instance, as part of its domestic military production, raised legitimate questions and serious concerns not only within the Middle East but also in the West. Of interest is Iran's possession of short-, medium-, and long-range missiles dubbed *Shahab* (Arabic for meteor) that can reach Tel Aviv.[35] I argue that, given the imposition of sanctions on Iran in the aftermath of the 1979 revolution, Iran decided to go it alone to develop its military capabilities. The production of guided missiles, tanks, unmanned aerial vehicles (UAVs), radar systems, and military vessels, among others, provides evidence of Iran building military assets in spite of sanctions.

It is imperative, however, to understand how military hardware is produced in Iran to have a better sense of how Iran's military development, however moderate it may be, has geopolitical implications. For instance, the production of Iranian guided missiles called "Dehlaviyeh," an advanced antiarmor system, is thought to be a replica of the Russian KBM 9M133 Kornet missile, according to military experts. The missile is believed to have been acquired not through procurement, as one might expect, but rather through Hezbollah, which in turn obtained it from Syria.[36] This came at a crucial time as the United States slapped more sanctions on Iran in the wake of its testing of long-range ballistic missiles, which violated a UN resolution.[37] In response to US sanctions, Iranian foreign minister Mohammad Javad Zarif argued that the Iran missile program is not covered under any provisions as part of the latest nuclear agreement signed between Iran and the P5+1. As a result, it did not violate United Nations Security Council's resolution 2231, signed in 2015.

Iran is communicating to the West that it has the capability and know-how to produce advanced systems through various means, including reverse engineering. The production of antitank, wire-guided US TOW-2 missiles (which were shipped to Hezbollah in Lebanon during its war against Israel in 2006) shows such knowledge at work. Similarly, according to analysts, Iran has recently produced various UAVs and appears to be perfecting the technology. That paves the way for Iran to produce other capabilities, such as aircraft and drones with stealth capabilities. Not surprisingly, the ability to produce and domestically develop UAVs with high-precision bombing strikes would put Iran ahead of other regional countries except Israel and Turkey. For example, two drone models Iran revealed (dubbed *Ra'd*, Arabic for thunder, and *Nazir*, Arabic for harbinger) would certainly change the balance of power.[38] On the topic of Iranian drones, do you recall the downing of the American ScanEagle, a short-range unmanned aircraft hovering over Iranian airspace in 2012? Iran provided Russia a copy of the drone as proof of its capability of producing high-caliber UAVs.

Iran's development of military hardware is not limited to missiles, tanks, and drones but also extends to advance-warning radar systems. For instance, Iran recently inaugurated two new radar systems: Arash-2 is capable of detecting small drones at a distance of about ninety-three miles, and Kayhan can detect cruise missiles.[39] I argue that the development of various military hardware foreshadows Iran's building far more advanced machinery now that sanctions are lifted. Nuclear capability is the ultimate objective.

Iran's Foreign Relations

I focus my narrative in this foreign-policy section on the developments and tenets on which Iran has structured its foreign policy from 1979 to the present. In subsequent chapters, I detail what Iran's foreign policy consists of and the strategies it pursues with countries inside and outside the greater Middle East.

Undoubtedly, the outcome of the 1979 revolution defines Iran's current foreign policy. At that time, Khomeini made sweeping changes to Iran's foreign-policy structure, changes that reversed course from policies the shah had implemented. Khomeini's justification for changing course was that the shah had invested himself and Iran too heavily in the West. The urgent need came to alert the masses to the potential risk of Western hegemony and thereby to reorient Iranian society.

This narrative fomented a new thinking within Iran's political elites and made it possible for the hard-liners, those in charge of the country, to ensure that Iran's new foreign-policy direction focused on four main elements. First, Iran needed to vehemently oppose the United States (which explains the slogan "Death to America" still in use today). Second, Iran needed to eliminate any outside influence in the country, a reference to the coup orchestrated by the Central Intelligence Agency that overthrew Mossadegh. Third, Iran saw the need to export its political Islam experience of government to other countries in the region and beyond in the hope of expanding the influence of its ideological principles. Fourth, the revolution of 1979 saw the establishment and rise to power of theocracy, an approach to governance that marginalized Iran. Thus, it sought to increase diplomatic engagement with developing countries.[40]

While the Islamic revolution produced religious and social tremors in the greater Middle East, countries like Lebanon and Syria welcomed the change. Others, including Saudi Arabia and other Gulf states, showed signs of political nervousness and fear over the possibility of Iran's revolution being exported, which could introduce defiance of their monarchies. However, Iran's new foreign-policy strategy also sought to calm the nerves of its neighbors by taking initiatives to improve relations. These steps were not enough to calm jittery neighboring states, who concluded that the success of the revolution could serve Iran's long-term objective to become the dominant power in the greater Middle East. Similarly, Iran worked discreetly to curtail the growing influence of the United States and other Western powers, especially after the 1973 oil embargo. Because of these dynamics, Iran started to build economic ties with other countries that share Iran's resentment of the West.

Make no mistake: Iran's foreign policy and relations with other nations have been tumultuous since the revolution of 1979 and the Iran hostage crisis, which deepened tensions between the United States and Iran. Further, the Iran-Iraq War reflected the animosity between not only those two countries but also among the other Muslim countries in the region, particularly how they perceived Iran in the postshah era. I detail the causes and geopolitical ramifications of the Iran-Iraq War in a separate section. That section highlights how the West realized that it was in its best interest to drag out the war to ensure that Iran stayed entangled in it for years.

Similarly, one must address Iran's foreign relations by emphasizing the concern and the fear other Middle Eastern countries had about the spread of Iran's Islamic revolution principles. In some ways, the fear is legitimate. However, the underlying issue had nothing to do with Iran. It had to do with the elites within those Arab countries fearing the sweeping changes within their own borders should a revolution reach their monarchies. Simply said, Iran's revolution demonstrated that power lies within the people, and it is the people's responsibility to ensure their proper governance. The difference in this case is that the revolution in Iran brought to the forefront the question of political Islam and whether the masses support adopting such a system.

While some Arab countries, mainly those in the Middle East, expressed concerns about the possibility of a geopolitical shift in the region because of the revolution, Iran had already embarked on a strategy to undermine certain regimes: "In 1981, Iran supported a plot to overthrow the Bahrain Government. In 1983, Iran expressed support for Shi'ites who bombed Western embassies in Kuwait, and in 1987, Iranian pilgrims rioted during the Hajj (pilgrimage) in Mecca, Saudi Arabia."[41]

Even greater Middle Eastern countries such as Egypt, Tunisia, and Algeria, countries with powerful fundamentalist movements, expressed suspicion and mistrust in Iran's experience with the revolution. Against these odds, Iran stayed the course, moving forward with its strategy of supporting proxies and groups including Hezbollah, Hamas, and the Palestinian Islamic Jihad, among others. While Iran pursued its strategy, countries like Egypt and Jordan established diplomatic ties with Israel. Iran and its proxies fiercely opposed the Arab-Israeli peace process.

Turning the conversation to address another continent, Iran's relations with Europe were marked by inconsistencies. Today, Europe sees economic opportunities in Iran after the lifting of sanctions. In 2016, the reality differs greatly from the postrevolution era that compelled European nations to reconsider their growing commercial ties with Iran given its human rights abuses and pursuit of nuclear program activities. This is no more evident than when President Khatami, a moderate, tried to reengage with Europe. While he was able to reset the tense relations, the election of his successor, Mahmoud Ahmadinejad, a hard-liner, shattered any hopes for the possibility of a constructive reengagement between Europe and Iran. Barbara Flanagan writes,

> President Khatami's efforts at engagement also went beyond the United States. He also tried to improve relations with many European states. While president, he embarked on a series of diplomatic visits to European countries, including Germany, France and Italy. This rapprochement was necessary after the assassinations in Europe. The European Union tried to strengthen Khatami, and the moderate reformists around him, through a policy of constructive dialogue in the later 1990s. Although Tehran saw improved economic relations with Europe during Khatami's presidency, President Ahmadinejad stepped back from some of Khatami's efforts. He placed more emphasis on the non-Western world, and concentrated more efforts on Latin and South America, including Venezuela, Brazil, and China.[42]

Those dynamics compelled Iran to reconsider its options. Iran chose to establish economic and military ties with Russia. Yet, many security analysts still question the meaning of that relationship or rapprochement. In my opinion, it is explained through the historical background that shaped Russia and Iran's relations over the years to what they are today. As of this writing, the rapprochement between Iran and Russia serves both countries' interests and furthers their own objectives, despite opposite trajectories. It also serves a far greater strategic aim: challenging the United States in view of its declining leadership in the Middle East. Yet, relations between Iran and Russia appear, at times, as though the countries have forged a strategic partnership; at other times, they seem to be far at odds. This puzzling relationship traces its roots to 1829, when Russia and Iran signed the Treaty of Turkmenchay, in which Iran lost to Russia the territory of modern-day Azerbaijan.[43] I believe that one should view Iran-Russia relations within this historical context, which left deep psychological scars and a sense of betrayal among Iranians, especially regarding their perception of Russia. That negative perception was reinforced when Russia recognized and supported Iran's Tudeh Party (translated as Party of the Masses of Iran), a communist party formed in 1941. Soleiman Mohsen Eskandari was the leader of that party. Less widely known is that the Tudeh Party played a pivotal role during the era of Mohammad Mossadegh when the latter decided to nationalize the Anglo-Iranian Oil Company, a move that on August 19, 1953, led to his ouster in a coup orchestrated by both the Central Intelligence Agency and the British MI6.[44]

Relations between Russia and Iran went through different stages that reflect-ed the changing geopolitical landscape. Khomeini's return from exile and acqui-sition of the helm of power in the Islamic Republic of Iran evidences one stage. His ascendancy to power forced the relationship between Russia and Iran to move from a territorial dispute to an ideological one. Russia, known then as the USSR, tried to influence Iran and pull it into its communist orbit in the Middle East, but it found that objective challenging. The rise of Khomeini-led political Islam proved difficult to influence and too strong to overcome. A large segment of the Iranian population who felt lied to and humiliated during the shah era found solace in the new political Islamic system brought by the revolution. The system allowed Islamic law to govern them, a powerful tool that cemented Khomeini's power over all sectors of the government. As a result, the idea of organizing op-position parties, either by the United States or by Russia, was doomed to fail. In fact, Russia (USSR back then) attempted to do exactly that but failed miserably.[45]

The complicated history between Iran and Russia has come full circle. Re-cent events have compelled both countries to set aside their differences and focus on a common goal: teaming up against the predatory forces of the West and the hegemony of the United States. Using history as my guide, I share the opinion of many security analysts that the political and economic collapse of the USSR exposed the weakness of Russia's political system and economic policies. The fall of the USSR diminished Russia's standing on the world stage and resulted in its loss of prestige. Searching for ways to revive its comatose economy and paralyzed political system, Russia saw in Iran an excellent opportunity. Russia's investment of time and resources in Iran proved wise. Now both Iran and Russia approach the world stage from two separate trajectories. I delve deeper into this topic in subsequent chapters when addressing Iran's nuclear program and the interna-tional community's position on the issue.

The State of the Economy

A wide consensus exists among scholars, analysts, and pundits that Iran's economy before the 1979 revolution held greater opportunities than after the revolution. As an agricultural society, Iran's economy grew rapidly before 1979. Development owed mainly to foreign investments and major industrialization in most econom-ic sectors. The energy industry took the lead in this development, and the flow of foreign investments, Western technology, and expertise supported it.

However, after the 1979 revolution, the economic settings turned upside down. The new government embarked on major economic reforms in response to the economic sanctions the West imposed on Iran. Those reforms included na-tionalizing the oil and financial industries. The government divided the economy into three separate sections: state, cooperative, and private. The supreme leader took sole control of the economic decision-making process. Lastly, the government took control in setting prices. Those measures have negatively impacted the economy, but not to the extent the West portrays it. Despite the sanctions, Iran managed to keep its economy afloat.

The big loser is the United States, which lost about $175 billion in potential export revenues to Iran.[46] I am interested to see whether the lifting of sanctions against Iran compels the latter to engage the United States economically or just maintain its current economic strategy of engaging Europe, Russia, and China. Early indications suggest that Iran has no intention, thus far, of engaging the United States economically. Iran has geared most, if not all, of its business deals toward the European Union, Russia, and China.

In the aftermath of sanctions, Iran's postrevolutionary government forged a path toward economic independence. It also tried to improve its citizens' standard of living. While the postrevolutionary-era government wants to immediately cement its authority over the economy, it fears civil unrest if, for instance, inflation becomes too high, prices continue to increase, wages stagnate, or the unemployment rate rises. A popular uprising would deliver a big blow to the revolution. It would cast doubt on the government's ability to improve economic conditions. Such an outcome would fuel more demonstrations in the country, thus calling for a return to the pre-1979 economic era and its conditions.

Government changes came as a demographic boom saw the population of Iran double between 1980 and 2000.[47] As a result, Iran's society became younger. This demographic change presented the central government and the supreme leader with major social challenges that require practical solutions more so than religious sermons and flowery speeches. In the 1980s came the bloody eight-year war with Iraq, a war that claimed at least three hundred thousand lives and cost Iran's economy approximately $500 billion.[48]

One must understand the historical context for the war between the two neighboring countries. To some, the war contributed another chapter in the ongoing saga of conflicts between the Persians and Arabs. To others, the war reflected not only the potential geopolitical changes in the Middle East because of the Islamic revolution in Iran but also Saddam Hussein's fear of such changes. Given that Iraq's majority are Shiites, Hussein feared that Iran's revolution could threaten the delicate balance between minority Sunnis and majority Shia in his country. Further, Iran's direct access to the Persian Gulf and Strait of Hormuz magnified Iraq's geostrategic vulnerabilities. One day, Saddam Hussein reasoned, Iran would exploit such geographic advantages. The Iran-Iraq War profoundly altered political trajectories in the Middle East. Chapter 6 in this book details the geopolitical implications now manifesting themselves in the region's political landscape. The invasion of Iraq, the civil wars in Syria and Yemen, the emergence of ISIS, the failed state in Libya, the political and security instability in Egypt, and the rift between the Kingdom of Saudi Arabia and the Islamic

Republic of Iran are cases in point. For now, suffice it to say the Iran-Iraq War had a major impact on Iran's economy.

We cannot talk about Iran's economy without highlighting the energy sector, mainly oil and natural gas, two major sources of Iran's revenues. In the first years after the revolution, sanctions against Iran coupled with strong oil output in the global market lessened Iran's economic output. Those factors compelled the Iranian government to adjust to new economic realities as sanctions bit hard into its economy. One must understand that the United States imposed no sanctions on Iran after the revolution until US embassy personnel in Tehran were taken hostage. During both the Carter and Reagan administrations, US sanctions on Iran limited the latter's ability to modernize its economic sector and update its outdated energy infrastructure. It was also during this era that the United States banned Iran's oil exports and froze the Iranian government's assets, totaling $12 billion.[49] Much of the 1980s saw a total embargo on Iran's trade and goods.

Similarly, American president Ronald Reagan pursued a policy like that of his predecessor, Jimmy Carter. However, Reagan decided to tighten the sanctions on Iran in response to regional events such as the Iran-Iraq War. Patrick Clawson, director of research at the Washington Institute for Near East Policy at the United States Institute of Peace, argues that the Reagan administration tightened sanctions on Iran during its eight-year war with Iraq to prevent the former from acquiring any dual-use items that could be modified for military use.[50]

However, a consensus among scholars suggests that, despite the sanctions, Iran managed to live within its means. Accordingly, Iran adjusted its economic output. Despite the embargo on its oil exports, most statistics point out that Iran's earnings from oil accounted for 58 percent while taxes represented 25 percent, and the Iranian government's other revenue sources counted for 17 percent of total national revenue.[51]

Despite the economic setbacks from ongoing sanctions, Iran marched forward with its market reform plans and industrial diversification. For instance, Iran began to support the pharmaceutical, biotechnology, and nanotechnology industries, fueled by a well-educated population. Many have asked in the wake of the lifting of sanctions as part of the 2015 nuclear agreement with the West whether Iran's reintegration in the world economy will transform it into an economic powerhouse. The answer is complicated given that it will take time for Iran to repair the economic damage the sanctions caused.[52]

Now that the sanctions on Iran are officially lifted and the gates to investments wide open, will Iran play a pivotal role in the world economy? Will its economic reintegration in the world translate into military and ideological opportunities? Will the expansion of Iran's political influence convince Russia and China to enter into a strategic alliance with Iran? Recently, current and former heads of state of major world economies, such as Germany, China, France, and Russia, have visited Tehran to secure their economic interests. I predict the trend will continue as Russia and China take advantage of the leadership vacuum resulting from the United States' leadership decline. Iran realizes that it makes sense to marginalize US economic interests in the region by not engaging American corporations when conducting major business projects. Iran's recent business agreement with a European consortium, rather than Boeing, to purchase 118 planes bears out this fact.

Iran's Tourism Industry

Iran is now witnessing a jump in the percentage of tourists visiting its famous cities to marvel at its ancient history and civilization. However, it took Iran a few years to rebound from the challenges its tourism industry experienced, mainly during its eight-year war with Iraq. The tourism industry has recovered slightly, providing Iran with revenue. Of interest in this conversation is how the tourism industry in Iran can potentially generate serious hard-currency revenues that beat its current revenues from exporting petroleum derivatives.[53] While Iran's historical sites and attractions have been compared to those of neighboring countries like Turkey, the consensus is that Iran will not see the same number of tourists that visit Turkey. The argument is that Iran's cleric-led government views certain aspects of tourism negatively. In other words, tourism may negatively influence its social and cultural values. Secularists and academics within Iran vehemently argue against this assertion. However, the thinking of Iran's leadership reflects its strong desire to preserve Iran's religious identity, which has come to define it since the 1979 revolution. Still, Iran is considered among the top ten countries in the Middle East in terms of tourism attraction, and the country has great potential for attracting international tourism.[54] Thus, despite resistance from its religious leaders, Iran is moving forward to diversify its revenue sources, hedging its bets that an expansion of its tourism sector will translate into more job opportunities.

Among Iran's most visited cities are Esfahan, Mashhad, Shiraz, and Tehran. Tourists include Europeans, Muslims from Asia and Africa, and even visitors from within the Middle East. For instance, in the 1970s, several organized tours from European countries such as Germany, France, and others visited Iran's famous archeological and historical sites, including the ruins of Persepolis dating to 518 BC and the seventeenth-century monumental arcades of Imam's Square in Esfahan.[55]

Yet, many in the West have portrayed Iran negatively. While its political vision since the revolution has posed serious concerns on the global stage, Iran's population has different views on its place in the world. Most Iranian people want to integrate into the world economy, to be part of the "global village," to thrive, compete, and succeed. It is fair to say the Western media presents a different picture, one of deep contradictions within the country. Even American visitors to

Iran disagree with the Western media's negative portrayal of Iran. Those visitors were surprised to learn that Iranians know a lot about American baseball, pop culture, and other significant American cultural happenings. On his last visit to Iran, in July 2015, BBC world affairs editor John Simpson wrote,

> It has been hard, over the years, to explain to Western readers and viewers the deep contradictions of Iran, one of the world's least-reported-on major countries. The problem is that we think we know what the Islamic republic is all about. We see the pictures of black-robed demonstrators in the streets denouncing the West and all its works. We recall the former president, Mahmoud Ahmadinejad, with his unshaven face and simian eyes, and think that he speaks for an entire nation of extremists. We assume, therefore, that Iran's nuclear programme is intended to wipe out Israel and threaten Western interests. And, as a result, we get Iran wrong every time.[56]

The question now that the economic sanctions on Iran are lifted is whether the country will improve its tourism industry. Will foreign investments in the hotel industry, for instance, rush in? The main challenge for Iran's leadership is to facilitate such a transition. Many in the West are curious to learn about an ancient civilization that influenced the history of the human race in many ways. If Iran takes advantage of that curiosity, the tourism industry will flourish.

Iran's Energy Sector

The government of Iran depends heavily on revenues from the energy sector. Iran is the second-largest exporter of oil within the Organization of the Petroleum Exporting Countries (OPEC) and fourth-largest producer after Saudi Arabia, Russia, and the United States. Unfortunately for Iran's economy, the economic sanctions over the years have choked its oil industry. However, one must put Iran's oil output in the global market in perspective. While Iran holds about 10 percent of the world's proven oil reserves, the lifting of sanctions will allow it to immediately increase its production by five hundred thousand barrels a day and by one million barrels a day within a month.[57]

Similarly, Iran has the world's second-largest gas reserves (after Russia), amounting to 29,600 bcm (i.e., around 15 percent of the world's total gas reserves).[58] Why is this important? The answer lies in the possibility that rapprochement between Iran and Russia will allow them to control the energy market, at least provisionally. Both countries' proximity to the Caspian and Black Seas would facilitate such control. Said differently, with Iran's position as the second-largest oil exporter, the lifting of sanctions allows it to be perceived as an energy power since no restrictions prevent it from selling oil to any country. However, Iran's ability to control the energy market, plausible before the drop in oil prices in 2014, might prove to be very challenging given recent changes in the global oil market. A steep drop in oil prices and the international sanctions

placed on Russia after its annexation of Crimea have worsened the Russian economy. The challenge for Putin is that Russia might find itself dealing with discontent, or worse, civil unrest, if the current economic downturn endures. That
outcome would force Putin to come up with solutions not only to save the Russian economy but also to salvage his legacy. Howard Amos argues, "While there
are few signs of economic problems translating into political protest, a prolonged
downturn threatens to undermine the prestige of President Vladimir Putin, who
has presided over an almost unbroken period of rising living standards during
his 16 years at the top of Kremlin politics."[59]

Under no circumstances am I suggesting the global oil market will return
to its pre-2014 market price of $120 a barrel any time soon. The possibility of
Iran and Russia controlling major portions of the energy market must be shelved
for now. However, as Russia reasserts itself in the Middle East, Iran is already
increasing its influence at the outset of a sanction-free era. Iran has begun to
position itself as a major, powerful, and influential player in the energy sector.
The entanglement of Saudi Arabia in Yemen, a raging civil war in Syria that has
become an international conflict among major powers, and the ambiguous foreign policy of the United States in the region all factor into the possibility of Iran
becoming precisely the power it seeks to be.

Conversely, Iran improved its hydroelectric industry mainly because of sanctions imposed on it by Europe and the United States in 2010 that targeted the energy
sector in order to hinder Iran's progress: "The EU sanctions approved at the June
summit include a ban on new investment, technical assistance and technology
transfers to Iran's huge gas and oil industry, particularly for refining and liquefied
natural gas."[60] Iran's upgrade of its energy infrastructure is a response to its growing
population. Could this be the reason Iran has always claimed that its nuclear program is for civilian purposes? Possibly, though not exclusively. The nuclear power
plant at Bushehr, operational since 2011, is an exception. Of note, the Bushehr power
plant was the second one built in the Middle East after the Metsamor in Armenia.[61]

Today's picture of Iran differs from the one from the mid-2000s. Some aspects of the sanctions on Iran have helped it to gradually integrate into the world
economy and forge economic ties with Russia, China, and, to some degree, the
European Union. However, the implementation of a new round of sanctions by
the United States highlights the long way Iran still has to travel to integrate fully
within the global market. Of interest in this discussion is Iran's recent test of
precision-guided ballistic missiles capable of delivering a nuclear warhead. The
test clearly violates the United Nations Security Council's ban issued in October
2015 and demonstrates Iran's defiance on multiple fronts, thus putting the United
States in an awkward political position. On one hand, the United States lobbied
hard to lift the economic sanctions on Iran once it agreed to the terms of the
nuclear deal. On the other, while Europe, China, and Russia may enter into economic agreements with Iran, which they did, the United States and Iran remain

far apart. As a result, their economic and diplomatic cooperation is unlikely, at least for now. The recent sanctions prove this point. Joel Schectman writes, "As a result, the United States is imposing sanctions on individuals and companies working to advance Iran's ballistic missile programme. 'And we are going to remain vigilant about it. We're not going to waver in the defence of our security or that of our allies and partners,' Obama said in a televised statement on Sunday morning from the White House."[62]

In the next chapter, I provide a detailed account of how Iran's status as the second-largest oil producer within OPEC will undoubtedly leave its mark on the world economy with the lifting of sanctions. I also address how the rapprochement of Iran with its main nuclear program backers, Russia and China, will affect energy security—and geopolitics, for that matter.

In conclusion, addressing other areas of modern-day Iran, such as education, demographics, language, architecture, and art would make an excellent contribution to the body of knowledge and give more richness to this book. However, my focus remains on how I see a nuclear Iran changing the geopolitical landscape of the Middle East. Such a change would alter the course of history and reintroduce the multipolar system that dominated relations between East and West for almost fifty years during the twentieth century.

4 Political Landscape of the Middle East

As of this writing, the world celebrates—perhaps not in a positive way—the one-hundred-year anniversary of the Sykes-Picot Agreement, a secret treaty between France and Great Britain signed in 1916, during World War I, in which the two countries negotiated the partition of the defeated Ottoman Empire. As the world recalls that treaty, chaos in the Middle East continues to unfold. Global-affairs analysts, academics, and governments on both sides of the Atlantic wonder whether, given the ongoing geopolitical shifts in the region, the Sykes-Picot Agreement can still hold.

Like other global-affairs analysts, I ponder how the rise of new players in the Middle East and the return of old ones might lead to a collision. These dynamics bring to the fore the question of whether the established political order in the Middle East as we know it can be sustained. Hardly. Borders between Iraq and Syria, for instance, have blurred. Yemen and Libya are failed states. An ongoing internal power struggle unfolds within the Saudi royal family. Egypt is in a chaotic state from both a political and a security point of view. These developments come on the heels of (a) the reemergence of Iran on the regional and global stage after its nuclear agreement with the West evidenced by the increase influence in the region through its military and economic realms; (b) Russia's assertive foreign policy in the Middle East; (c) the cautious, discreet diplomacy of China; and (d) the ambiguous foreign policy of the United States in the Middle East. The aforementioned suggest that the postcolonial order is about to change. There are those like Jon B. Alterman, director of the Middle East Program at the Center for Strategic and International Studies in Washington, DC, who believe that, despite the chaos in the Middle East, the states where the conflicts are enduring will remain, though maybe under different circumstances. He writes,

> One hundred years into the Sykes-Picot Agreement, a growing chorus of voices is asserting its imminent demise. Skeptics say that few of the Arab states' borders ever made any sense, and the uprisings sweeping through Syria, Iraq, Libya, Yemen and elsewhere represent the long-overdue death rattle of the postcolonial order in the Middle East. Erasing states is not nearly as easy as it sounds, though, and proto-states that had only shallow rationales in 1916 have sunk deep roots in the century since. They cannot simply be swept away, and breaking them up will do little to promote the domestic harmony that they have failed to provide. When the dust settles, whenever that is, we are much more likely to see new kinds of states within the same borders that we see now in the Middle East than we are to see new states inside new borders.[1]

Addressing the political landscape of the Middle East is no easy task given the region's unpredictability, shifting loyalties, mentality of mistrust, and constant changes. However, I present a framework that puts this chapter, along with its sections, in perspective within the context of geostrategic interests. Doing so provides readers with a point of reference for acquiring a better understanding of how the current shift in geopolitics in the Middle East, which has been at work for the last thirty years or so, did not occur in a vacuum but rather is driven by several factors. That said, I do not intend to address the history of the early decades of the Middle East. Rather, I focus my narrative on the political, ideological, and religious happenings after World War II that marked the course of history in this troubled region.

Is it imperative to approach the chapter in this manner, and, if so, what is the wisdom in doing so? The answer lies in the content of one specific sentence: Events in the Middle East do not occur in a vacuum. An invisible hand works with hidden forces behind the scenes to create an agenda to follow, a policy to implement, interests to gain, and regimes—including totalitarian ones—to protect. As I stated earlier, it is a daunting task to cover the political landscape of the Middle East in a few pages or to decide on the historical frame of reference. Addressing, for instance, the contemporary political landscape of the Middle East and its involvement with China, Russia, and the United States would, now more than ever, make an excellent topic for a new book. That said, my main focus remains on how the political landscape in the region after World War II shaped, influenced, modified, and challenged all conventional geopolitical theories. Further, the political landscape challenges conventional assumptions, suggesting that major powers firmly control the Middle East. The latter assertion might hold some truth given the dominant role the British Empire played in the region before, during, and after World War I. Consider, for that matter, the central role the United States played in the region from World War II until about the end of the Cold War.

However, the current state of affairs in the Middle East put these assumptions under the microscope of reality. After all, empires do not last forever. The end of the British Empire's rule in the Middle East is a case in point. The new realities in the Middle East, including the roles Russia and China now play, roles facilitated through Iran's nuclear agreement with the West and the civil war in Syria, lend strength to my argument. Or consider Russia's sale of advanced weapons to the Kingdom of Saudi Arabia or the military coup in Egypt following the election of Mohamed Morsi of the Muslim Brotherhood, followed by an exchange of high-level visits between Moscow and Cairo. All these dynamics have a major impact on the geopolitical scenery of the Middle East, whether the West likes it or not.

My main focus remains on addressing specific events that I believe have shaped the region's political destiny into what it is today. These events differ in nature, outcome, and substance, yet they share a commonality, some sort of link that ties them together. That is because the Middle East is no ordinary region; it is a mixture of cultures, tribal mentalities, religious ideologies, political orientations, and geopolitical aspirations. To better understand the Middle East's geopolitical landscape, one should spend time there to fully appreciate its complexity. And, if the locals, who are familiar with the culture and all its aspects (language, traditions, and customs) cannot decipher the turbulent nature of the region, how would the major powers—the United States, Russia, China, and others—be able to? Such arrogance! After all, it is hubris to think that any particular country has figured out the exact formula consisting of all the right variables to solve the complicated issues of an ever-changing, unpredictable region. The Middle East will remain as complicated and volatile as ever.

To better understand the geopolitical shift in the Middle East, I provide a snapshot of the scale of the political and social movements that have influenced and continue to shape the region. Such a snapshot, at the same time, brings questions about the United States' influence or lack thereof, decline in leadership, and hegemony to the forefront. Recall when the Arab Spring erupted in Tunisia and engulfed more than twenty countries, including Egypt, Yemen, Jordan, Bahrain, Algeria, Libya, Saudi Arabia, and others. The resulting political quake rattled the political and social landscape of the Middle East and caught the United States off balance, more than any other major power. As the guarantor of peace and stability in the region, the United States was surprised by the sudden geopolitical shift. While the notion still holds that the United States is the only major power heavily involved and invested in the Middle East, the current reality on the ground suggests otherwise. China and Russia play far greater roles than they have since the end of the Cold War. To make my point, in the subsequent pages, I address particularly China's reintegration in the region and its economic and military motives.

The question I wish scholars and global-affairs analysts to address aggressively is, what sorts of US foreign-policy strategies have contributed to this turmoil? Is there a specific policy that we can pinpoint? Hardly. David Crist writes, "But the Iran policy, or lack of it, frustrated Myers. The 'axis of evil' line typified the problem within the administration in the chairman's mind. 'There were attempts to address Iran strategically,' Myers said, 'but at the NSC level, it always became too sensitive or not the right time.' Myers suspected there might be something going on behind the scenes, perhaps a backdoor strategy that he was not privy to; in reality, however, the Bush administration had developed no Iran strategy."[2]

Needless to say, many events in the Middle East, past and present, contributed to this shift in the geopolitical landscape of the region, thus complicating the United States' strategy toward the Middle East and rendering it ineffective. For instance, since the beginning of the Syrian war, the United States has been unable to formulate an effective strategy to the ongoing conflict there. Is it because the conflict proves too complicated for the United States to understand? Is it because the US intelligence apparatus is unable to penetrate fighting groups there? Or is it simply that the United States has no strategy in the first place?

Whatever the answer is, inaction, diplomatic and otherwise, sends a negative message to the region and thereby confirms the United States' decline in leadership in the Middle East and beyond. However, one must understand that the United States' involvement in the Middle East did not happen overnight and that finding solutions to these complex issues will take time, patience, and a well-defined strategy. Since the conclusion of World War II, US interests in the

Middle East have expanded exponentially. During World War II, competition over resources, mainly energy, was fierce. The USSR wanted to add some regional countries to its communist orbit. As a result, competition over dominance of the Middle East between the United States and the USSR was managed through the Cold War.

However, after the Soviet Union disintegrated, the United States found itself the sole superpower in the world. Hence, it had to bear the burden of world leadership on its shoulders. The Middle East has always played an important role in the United States' global strategic interests. These interests include 1) the flow of oil and gas to the world markets, 2) markets for the sale of American weaponry, and 3) access to international markets for US corporations.

The attacks of 9/11 changed the United States' foreign policy priorities: shifting from focusing on relations with major powers including Russia and China to an emphasis on rogue states and non-state terrorist groups such as Al Qaeda. The change in US strategy, manifested by a militarized foreign policy known as the "war on terror," placed US foreign policy under scrutiny. Questions as to the wisdom of this policy came to the fore. Make no mistake: the military's role in US foreign policy in the aftermath of 9/11 was both necessary and tragic. On the one hand, it was necessary because the United States had to act militarily against terrorist groups wherever they were. On the other hand, it was tragic because diplomacy took a backseat, and most, if not all, US foreign-policy engagements were approached from a military perspective.

And that is exactly what happened in the Middle East. The shift in policies toward the region profoundly and negatively impacted the United States' future there. Countries in the Middle East started to question whether they should support US efforts in the so-called war on terror. Most global think-tank analysts took notice of the United States' new strategy. For instance, Gao Zugui, an international political studies expert in China, argues, "Since the beginning of the [twenty-first] century, and especially after the 9/11 incident, America has paid more attention to the Middle East. During the [eight] years when George Bush junior was in office, America made the Middle East and even the 'greater Middle East' the priority in its national security strategy."[3]

As of this writing, the ongoing upheavals in the Middle East challenge Washington's efforts there. This is not just about the civil war in Syria and the chaos in Yemen and Libya but also the instability in Egypt, the Saudi royal family's internal feud, and Iran's growing military and economic influence in the region, among other issues. This turmoil undoubtedly undermines US strategic interests in the region.

The challenge for the Washington establishment is a turmoil not limited to one specific Arab country in the region but rather spread from one state to another. Yes, the Arab Spring erupted in Tunisia; yet, there were demonstrations in Saudi Arabia's eastern region, Kuwait, Iraq, Yemen, and Bahrain, among other places. These countries are considered US allies, and they are expected to support the United States' agenda in the region. Not so fast, I argue. While the Arab Spring succeeded in overthrowing several of the Middle East's longtime dictators in the hope of a better future, the West was astonished to realize that those efforts resulted in something it failed to factor in before providing covert support: the rise to power of Islamist groups.[4]

Turning my attention to Iran is therefore the logical approach given that Iran is not only a Middle Eastern country where the action is taking place but also one that is influencing and contributing to the shift in the region's geopolitical landscape. Of interest in this discussion is how Iran's strategic thinking derives from the current security environment in the Middle East. Past events—including the 1979 Islamic revolution, the Khobar Towers' attack in Saudi Arabia, the Iran-Iraq War, the 1973 oil embargo, the 1979 hostage crisis in Iran, and both Gulf Wars, among others—impacted US foreign policy toward the region. Similarly, recent events, such as the Iran nuclear agreement with the West, the sectarian violence in Iraq, the quagmire in Yemen, and so forth, define Iran's strategic framework, the one on which it can pursue its geopolitical aspirations in the region.

Of course, we have heard on many occasions that Iran is meddling into the internal affairs of some countries in the region in addition to expanding its military and ideological influence; the assertion is true. What we have *not* heard is that events like the Iran-Iraq War helped to convince Iran that it could indeed influence political outcomes in the region. The Washington establishment miscalculated in that matter. The assumption that American and European sanctions, for that matter, would limit Iran, restraining it from expanding its sphere of influence in the region, was overstated. On the contrary, the ongoing upheavals in the region and the United States' ill-conceived policies (its invasion of Iraq, its nuclear agreement with Iran, its turning a blind eye on the military coup in Egypt) allow Iran to focus more on where and how to devote its resources to further entangle the United States in the Middle East.

The bottom line is that the longer the conflicts and upheavals persist in the Middle East, the better it is for Iran, strategically speaking, to expand its influence. The matter is even more worrying now that Iran joins forces with Russia and China to further undermine the United States' leadership, whatever is left of it. Iran's rapprochement with Russia and China also communicates to the rest of the world that the Middle East's political order is shifting, whether the West (the United States included) likes it or not. The contribution to this shift is not conducted unilaterally (Iran) but rather multilaterally (Russia, China, India, and Iraq, to some degree).

In support of the latter assertion, consider how Iran was able to influence and sabotage any policy the United States tried to implement in the region. One does not have to look far to see how Iran uses its agents, be they in Iraq (Mehdi Army during the US invasion of Iraq in 2003), Lebanon (Hezbollah during the 2006 war against Israel), Yemen (Houthis against the Yemeni government supported by Saudi Arabia), and even Palestine (Hamas against Israel). Equally important, Iran's influence manifests itself in the Gulf states, such as Bahrain, where demonstrations as part of the Arab Spring fueled speculations that Iran is behind the rise of the Shiites there.

But let us assume for a moment that one does not believe in Iran's ability to destabilize the region through its support of various proxies. What can the United States do to limit Iran's influence? Nothing. It is not because the United States does not want to, but because it *cannot*. Since the latter assertion is one of the themes of this book, I address it in detail in chapter 6. For now, suffice it to say that all these developments on the political stage of the Middle East, with Iran's behavior at its center, trace their origins to events from decades ago. This poses a tremendous challenge for the United States as it tries to figure out what its policy, if any, toward the region is or ought to be.

The United States' effort to develop a foreign policy in the Middle East is complicated by the presence of several countries on the political landscape of the Middle East now trying to influence outcomes to their favor. On the one hand, one finds Russia, China, and Iran. On the other hand, one has the United States and its traditional Western allies, which include the United Kingdom, France, and Germany. The region's current political environment introduces other players that can no longer be ignored: Iran, Saudi Arabia, and Turkey. Each country wants to ensure that the outcome of the ongoing conflicts works to its favor. The civil wars in Syria and Yemen provide examples of this assertion.

Iran-Iraq War

I address the Iran-Iraq War here not only to delve deeper into the root causes of the conflict itself, but also to present the conflict from both Iraq's and Iran's perspective. The overall objective of this section is to show how the conflict shaped, and continues to shape, Iran's political engagement on domestic and global issues. Further, I provide a framework that puts Iran's nuclear ambitions under the lens of reality, a reality influenced by not only the 1979 revolution's outcome but also its geopolitical impact.

Brief History of the Conflict

If we ask, "What was the purpose of the Iran-Iraq War?" answers will vary depending on whom we ask. Yet, most historians agree that geographic disagreement over Shatt al-'Arab river and other disputed territories—including the three islands near the Hormuz Strait—was the reason for the war. These areas have a profound economic and military importance for both countries, yet neither country proved competent in its approach to the bloody war. The result was an inconclusive struggle that ended with devastation, collapse of both contestants, and lack of a victor. Nevertheless, before any of that, one must understand the root causes and the impact the eight-year war had not only on Iran and Iraq but also the region and global politics overall.

While I believe the conflict is rooted in three different components—territorial, ideological, and personal—the consensus is that territory played the greatest role in defining the conflict, since both countries were known for having major underground oil reserves. The territorial dimension is difficult to cover in a few pages and frankly goes beyond the scope of this narrative. One must understand, however, that control over the territory of Shatt al-Arab region was the source of the conflict. As one who is intimate with the culture of the Middle East, I believe the defining elements of the conflict rest on what separates Iran and Iraq linguistically, culturally, and ideologically. They are two different countries with two different cultures: Iraq is Arab, Iran is Persian.

The turbulent history, even animosity, between the two cultures traces its roots to the seventh century when the Islamic Empire defeated the great Persian Empire. As a result, the Arabs absorbed both the Persian lands and population. While Persians embraced Islam through conversion, they did not want to give up their Persian identity, language, and culture. Even today, some in the West who have no historical knowledge of the region still think the Iranians are Arab rather than Persian. They are not. Tensions during the seventh century made possible the rise of the Shiite sect.

Fast-forward to the nineteenth century, which saw British domination of the region. The Brits competed with the Russian Empire alone for regional control. The rivalry got even tenser after the discovery of oil. I strongly believe the outcome would have been different if not for World War I, which turned their rivalry into cooperation as both empires, the Russian and the British, realized what they could achieve if they defeated the Ottoman Empire. Embarking on that adventure and defeating the Turks cleared the way for the British Empire to control both Iran and Iraq.[5] A series of treaties followed between Iran and Iraq until the 1979 revolution that saw the ouster of the shah. The revolution threw Iran into chaos, and the military establishment became fragmented, disorganized, and weak. The outcome provided Iraq a historic opportunity to settle not only old scores but also the border dispute once and for all. Saddam Hussein's thinking proved to be a strategic mistake of historic proportions. The outcome of the Iran-Iraq War made the Middle East what it is today.

While the territorial dispute was at the center of the conflict, one cannot ignore the role ideology played in shaping its outcome. Of interest is how the countries were ideologically opposed: "The two warring countries are controlled by regimes that hold diametrically opposed ideologies. While Iran's regime is theocratic and Universalist, Iraq's is secularist and nationalist,"[6] Suleiman Kassicieh, professor at the University of New Mexico, argues. The conflict cost both nations staggering amounts of money and priceless human lives. It destroyed economic growth for both Iran and Iraq. Further, economic loss extended beyond the borders of both countries to include the Gulf region, sending political tremors throughout the countries of the Gulf Cooperation Council.

Many historians agree that the Iran-Iraq conflict erupted when Saddam Hussein invaded Iran in September 1980. Most agree that Hussein feared that the Islamic revolution of 1979 might spread to his ethnically diverse country of Iraq, thus inspiring the majority Shia to revolt against his rule. As a result, scholars argue that Hussein's fear of an exported Islamic revolution played a major role in the tense history between Iraq and Iran. The 1979 revolution in Iran created chaos and threw the country into anarchy; Hussein, thinking he could take advantage of the chaos, made a strategic mistake by attacking Iran. He discovered that the attack allowed Iranians of all stripes to join in defense of their country. By 1982, Iran had regained some of its lost territory. This conflict marked the beginning of the involvement of different players, regional and global, that shaped the outcome and changed the region's geopolitical calculations.

Before proceeding further, I must address the conflict from the Iraqi and Iranian perspective separately. These vantage points give readers a clear view of how each country argued, justified, and managed expectations, domestically and globally.

Iraq's Perspective

Saddam Hussein ascended to power in 1979, creating anxiety in the Middle East as Iran's Islamic revolution was unfolding. Simultaneously, questions arose about his regime's legitimacy. These questions, at first political issues, became military concerns.[7] While such questions swirled in the region, other global developments gave context to happenings in the Middle East: the fall of Phnom Penh and the collapse of the Pol Pot regime; the nuclear plant accident at Three-Mile Island in Pennsylvania; the election of Margaret Thatcher as prime minister of the United Kingdom; the agreement between the United States and USSR to sign the treaties proposed at the Strategic Arms Limitation Talks; Iranian militants' seizure of the US embassy in Tehran and hostage taking; and, lastly, the USSR's invasion of Afghanistan. The 1979 Islamic revolution sent shock waves throughout not only the region but also the world.

Two important points follow. First, the overthrow of regimes like the ones in Iran (the Pahlavi monarchy) and in West Asia, including Syria, in 1949 rendered the United States helpless in suppressing the massive popular uprising. Thus, the United States and its allies went into political panic mode. Second, changes on the Iraqi political landscape brought with them major political and economic transformations. For instance, Saddam Hussein used oil wealth to revamp the Iraqi military apparatus. That initiative was evidenced by massive increases in military expenditures, which reached about 8.4 percent of gross national product in 1979.[8] One must understand, however, that the need for the military buildup

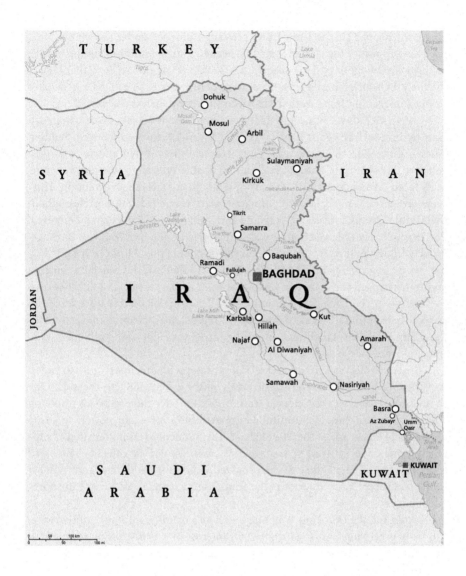

traces its roots to 1958, when Iraq became a major military market for the USSR as it pulled Iraq into its communist orbit.

In this era, the United States became more interested in the Middle East. Following the wane of British power in the region, competition between the United States and Britain became evident. However, the United States and the United Kingdom agreed to share in developing the Middle East's rich resources, including oil.[9] In 1972, Iraq signed a military treaty, among other cooperative treaties, with the USSR. Thus, Iraq was set to develop nuclear capabilities. Saddam

Hussein reasoned that to limit Iran's growing influence under the shah regime with the support of the United States, he needed a strong military equipped with sophisticated weaponry. However, Iran's Islamic revolution in 1979 that ousted the shah, a US ally, relieved Hussein of some of his worries since the United States stopped upgrading Iran's military hardware after the shah was deposed.

Conventional wisdom suggests that more than one specific issue compelled Iraq to invade Iran. Kurt Burch writes, "Several factors influenced Saddam Hussein's decision to attack Iran. First, Hussein's iron-fisted efforts to forge a cohesive, secular Iraqi nation-state among hostile ethnic groups and religious sects faced tremendous challenges from Iran's fundamentalist revolution. Hussein viewed Iranian officials' inflammatory plans to incite revolution throughout the Middle East as a threat to Arab and Iraqi influence in the Persian Gulf area. For example, Iran's leaders and its Shia majority population touted a 'brotherhood' with the politically repressed but majority Shia population in Iraq."[10] Under the prism of geopolitical realities of the Middle East, the war between Iraq and Iran changed the course of history, at least in the region, and affected relations between the United States and USSR. Thus, major powers were forced to rework their geopolitical calculations in the region and reposition themselves for the emerging political realities that resulted from the eight-year conflict between Iraq and Iran.

The Iran-Iraq War provided the United States with an opportunity to widen its sphere of influence in the region while pushing the USSR to the side. One should note that Iraq's invasion of Iran in 1980—under the pretext of resolving a border dispute—provided unlimited opportunities for major powers to shape the political landscape of the Middle East. This influence was not limited to supporting one country (Iran or Iraq, in this case) against the other but involved supporting both. The United States provided weapons to both Iraq and Iran, for example. Of particular interest is the Iran-Contra scandal, addressed in subsequent pages.

Regarding the Iran-Iraq War, one sees how a conflict, military or otherwise, can truly precipitate shifting loyalties. Of interest in this discussion is how Iraq forsook its ties with the USSR and instead realigned itself with the United States. Unknown to many at that time, talks had already taken place behind closed doors on reestablishing diplomatic ties between the United States and Iraq. Those talks avoided making any direct link regarding the real motives for the United States supporting Iraq during its eight-year war with Iran. Could this explain the sudden removal of Iraq in 1982 from the US State Department list of "state sponsors of terrorism"? The facts point to that outcome: "As Saddam Hussein later revealed, the United States and Iraq decided to re-establish diplomatic relations—broken off after the 1967 war with Israel—just before Iraq's invasion of Iran in 1980 (the actual implementation was delayed for a few more years in order not to make the linkage too explicit). Diplomatic relations between the US and Iraq were formally restored in 1984—well after the US knew, and a UN team confirmed, that Iraq was using chemical weapons against the Iranian troops."[11]

The Iran-Iraq War provided economic opportunities for all interested parties. For instance, France and Great Britain were major sources of weapons for Iraq. Equally important, some of Iraq's uranium used during the conflict originated in Portugal, Italy, and France. What remained undisclosed is how not only major powers but also other countries profited from the Iran-Iraq War. It benefited many countries for the conflict to drag on. Glenn Frankel, author and former editor of the *Washington Post*, argues, "Everyone, it seems, took a slice of the Iraqi arms pie. The Soviet Union, France, China, and Chile sold Baghdad much of its off-the-shelf weaponry. West Germany, France, Britain, the United States, Belgium, Austria, Switzerland, and Brazil all sold the components, machines and tools—much of it material with civilian as well as military application—that are the building blocks of the modern Iraqi war machine."[12]

Similarly, the United States was the major driving force influencing the outcome of the Iran-Iraq War to its favor. As a result, the United States was willing to do whatever it took to ensure that Iraq won the war. For instance, the United States provided "crop-spraying" helicopters that Iraq used in its chemical attacks against Iran. There is, however, reference in the literature to the helicopters the United States sold Iraq having both civilian and military uses. Norm Dixon writes, "Conventional military sales resumed in December 1982. In 1983, the Reagan administration approved the sale of 60 Hughes helicopters to Iraq in 1983 'for civilian use.' However, as Phythian pointed out, these aircraft could be 'weaponised' within hours of delivery. Then US Secretary of State George Schultz and commerce secretary George Baldridge also lobbied for the delivery of Bell helicopters equipped for 'crop spraying.' It is believed that US-supplied choppers were used in the 1988 chemical attack on the Kurdish village of Halabja, which killed 5,000 people."[13]

My focus in this section is not solely on the role the United States played in influencing the outcome of the Iran-Iraq conflict to suit its strategic interests. Nor is the emphasis on how the war itself came to define Iran and Iraq, politically and otherwise. Rather, the important fact that remains unaddressed, conveniently or ignorantly, was the extent of the United States' involvement in Iraq's "weapons of mass destruction" program, which became public knowledge in 2002. During this time frame, Iraq was prepared to turn over a lengthy report (about 11,800 pages) regarding its weapons of mass destruction programs.[14] Unknown to the rest of the world at that time, the United States examined the report before its submission to the United Nations Security Council and removed about eight thousand pages that could have revealed the direct involvement of US companies in this program. Iraq nevertheless disclosed the contents of those pages, revealing to nonpermanent members of the UN Security Council the involvement of American firms including Honeywell, Semetex, Rockwell, DuPont, and

Bechtel, among others. The disclosure also showed that Germany supplied Iraq with technology and military equipment. These revelations appeared in the German magazine *Die Tageszeitung* (Berlin Daily), on December 18, 2002.[15]

Why is the disclosure of this information crucial? The answer lies in understanding what happens later, especially with the United States' invasion of Iraq in 2003. One concludes that there is no doubt that the United States and other countries benefited from the Iran-Iraq War, which shaped not only relations among different countries and impacted tensions within the region but also framed future relations among global powers.

Iran's Perspective

The Iran-Iraq War meant more than just military engagement. It revolutionized the thinking of Iranian society. It shaped Iran's ideological aspirations as part of the 1979 revolution's principles. The war helped Iran write its own history, and the West did not see it coming since it assumed that Iraq would defeat Iran. Finally, the war served as a platform on which Iran reaffirmed and launched its religious, political, and economic aspirations since it proved how resilient it can be in the face of hostilities, including war. Lifting sanctions on Iran in the aftermath of the nuclear agreement with the West testifies to this fact. The Iran-Iraq War paved the way for the Islamic republic to position itself for this outcome.

Why does the grueling war with Iraq from 1980 to 1988 continue to shape the thinking of Iran's elite and the fabric of Iranian society and yet get little to no attention in the West, including in the United States? Ask Americans, at least those who care to read history and learn from global events, what first comes to mind when Iran is mentioned. The likely answer is "the CIA's coup against Mohammad Mossadegh," or "the 1979 Islamic revolution," or the "US embassy hostage crisis in Tehran." Saddam Hussein's September 22, 1980, invasion of Iran is not high on the list of the US foreign-policy establishment or those interested in Iran. Yet, the Iran-Iraq War is an ideological, economic, political, religious, military, security, and geopolitical story. The Iran-Iraq War is too important for people interested in the Middle East to ignore.

In my opinion, there are four major reasons why the Iran-Iraq War still matters to Iran and the region, but, unfortunately, not to the West. First, the conflict mirrors the ongoing ideology of Iran's 1979 revolution. Said differently, after the shah's overthrow and the ascent to power of Ayatollah Ruhollah Khomeini, the elite argued that the conflict with Iraq must be seen within a religious framework—as a holy war of some sort. Such a portrayal not only plays well with Iranians of all stripes but also helps recruit fighters. As a result, the revolutionary elites headed by Khomeini saw an opportunity to use the ideology (the principles of the revolution) wrapped in a religious narrative that blindly removed boundaries within society, and they rallied behind the leadership.

Behnam Ben Taleblu, an Iranian research analyst at the Foundation for Defense of Democracies, argues, "For Iran's leaders, the ideological magnitude of the war helped blur national boundaries, parsimoniously dividing the world into good and evil."[16] Khomeini cleverly exploited the emotional vulnerabilities of an Iranian society that had suffered tremendously, in his view, at the hands of the Americans. As a result, it made sense to him to use the same rhetoric and anti-United States slogans against Saddam Hussein. In that way, the Iranians justified their support for Khomeini and defended Iran at all costs. That explains why Iraq and the West were astonished to learn how the conflict rallied all Iranians to strongly resist foreign forces, military and ideological.

The issue does not end there. The conflict provided Khomeini a platform on which to export his religious, revolutionary ideology to other parts of the Muslim world, arguing that the Iraqi people must challenge their leadership (a reference to Hussein) who had now risen against Islam. As I argued in my previous book, *The Ambiguous Foreign Policy of the United States toward the Muslim World: More than a Handshake*, Islam has often been used to legitimate whatever agenda the leadership wants to advance. In the example of the Iran-Iraq War, Khomeini wanted to reach different corners of the Muslim world, and, to some degree, he has succeeded in his objective. The emergence of the Lebanese Hezbollah exemplifies this point.

The bottom line is that the Iran-Iraq conflict served two purposes for Iran: (1) it afforded Khomeini the opportunity to export his revolution and reach different parts of the Muslim world; and (2) the conflict, at least from Iran's perspective, sent a message to other countries to think twice before attacking Iran. In my opinion, there is one caveat: the only difference between the outcome of the war and the political shift in the region is that Iran never limited the spread of its revolution to the concepts and theories Khomeini preached. Rather, it manifested itself through the barrels of AK-47s and bombs. This explains why Iran's domestic politics took priority over the manner in which the war was prosecuted, characterized in the slogan "Revolution before Victory."[17]

Not surprisingly, those who followed Khomeini championed his ideology and referred to the Iran-Iraq War as the "Holy Defense." Iranian leaders who participated in the war and now happen to hold positions in the government blame the United States for the conflict. Even Iran's current supreme leader, Ayatollah Ali Khamenei, argues that the United States gave the green light to Iraq to attack Iran.

The memory of the Iran-Iraq War will live on for generations. Iranian leaders constantly refer to the conflict in their speeches to remind Iranians of their fellow citizens' sacrifices, their triumphs, and their leadership's fidelity to the virtues, concepts, and time-tested principles of the Islamic revolution.

The second reason why the Iran-Iraq War still matters to Iran and the region is that the conflict carried significant political meaning wrapped in a religious narrative. As the war took on a life of its own, the charismatic Ayatollah Khomeini wanted to ensure that the legacy of his revolution would live on and motivate the nation to stand behind him. He tapped into the nation's emotions by calling on the people of Iran to continue their sacrifices for the sake of Iranian identity, for the spirit and benefit of the revolution, and for the defense of the religion of Islam.

Make no mistake: not everyone was committed to Ayatollah Khomeini's revolutionary vision and approach to the conflict. Yet, Khomeini's ability to tap into the mind set of Iranian society using the Iran-Iraq conflict as a platform on which to marginalize his opponents served him well on two fronts. First, by marginalizing his opponents, Khomeini identified those who might sabotage his plan and overall strategy (exporting the Islamic revolution to the Muslim world) years down the road. Second, by identifying his opponents, he was able to tell the rest of society to look at how agents of the West were trying to undermine the revolution and destroy the will of Iranians to stand up to the West and its agents, including Saddam Hussein. I argue that the conflict provided Khomeini legitimacy and political capital at the domestic level. The conflict also sent a message to the rest of the world that Iran would not return to the era of the shah.

While he infused his rhetoric and fiery speeches with a religious tone, Khomeini realized the tremendous destruction and loss of lives his country suffered. Could this explain why current Iranian leaders, the mullahs, are constantly trying to revive the spirit of the revolution while instilling a culture of sacrifice and martyrdom in a society that has outgrown it? The answer is yes, as the generational gap continues to widen. This is evidenced through statements the Iranian leadership issues when referring to the "Holy Defense." Consider the following: "Faced with generational changes, Iranian leaders, such as Armed Forces Chief of Staff Major General Seyyed Hassan Firouzabadi, continue to praise the war, noting that 'the Holy Defense, much like the Islamic Revolution, must forever remain alive.' Iran has obliged, creating museums that enshrine the sacrifices of those who partook in the war. Cognizant of the diverse and sharply differing views of Iran's populace, and especially its youth, officials have sought to use the Iran-Iraq War to promote 'unity,' while extolling the virtues of knowing the war 'for the postwar generation.'"[18] It is fair to say that Khomeini's use of the conflict to create a culture of sacrifice served him well during his tenure. Ultimately, Khomeini's beliefs became the platform on which Iran coped with a plethora of issues following his death. Of interest in this discussion is how Iranian leaders approached the sanctions, for instance.

Third, Ayatollah Khomeini used the conflict as a conduit to disseminate his agenda: to export the beliefs of his Islamic revolution to the rest of the Muslim world. To achieve this, it was paramount for Khomeini to promote the conflict, from a security perceptive, as both holy (on a religious ground) and defensive (on a militaristic ground). As devastating as the war was, it helped Iran to objectively assess its military vulnerabilities. The lessons Iran learned from its war with Iraq include the importance of ballistic missiles and the significance of biochemical weapons—and the need for antishipping missiles, thus, paved the way for it to develop a strategy by improving its domestic military capabilities to defend itself against any future threats. Today, Iran has an expansive ballistic missile arsenal, asymmetric naval warfare, and ties to nonstate militant groups.

That image of valor also served as a deterrent, making its enemies think twice before engaging it militarily.

Once again, I find myself wondering, could this demeanor be what made the West refrain from attacking Iran's nuclear facilities before reaching a nuclear agreement? I think so. The West had to think twice about the consequences of attacking Iran. Obviously, having analyzed the various scenarios that might result from attacking Iran, the West disliked the plausible outcomes. The thinking among the Iranian leadership is best exemplified by one of the Supreme Cultural Revolution Council members, who stated, "If it wasn't for this eight years of war, ten wars would be imposed on us. And these very same wars that they have commenced in Lebanon, Syria and Iraq, they would have created in Iran. . . . And it is the Holy Defense that has made the enemy not dare attack us."[19]

The conflict motivated Iran to adjust and reorganize its military apparatus. Viewing the conflict through a strategic lens afforded Iran the forward thinking it needed to focus its strategic vision on improving its military infrastructure and learning how to be self-sufficient as sanctions started to bite into its economy. All these dynamics prompted Iran to invest in its weak infrastructure and military apparatus. Take, for instance, the long-term impact of the foresight of the Iranian Revolutionary Guard Corps officer who saw an opportunity in "reverse engineering," which became the basis of Iran's missile technology program. Behnam Taleblu, a nonresident research fellow at the Foundation for Defense of Democracies, writes, "In time, Iran moved to invest in a host of unconventional capabilities and asymmetric tools. According to UNSCOM [United Nations Special Commission] figures, roughly 63 percent of Saddam's missiles were fired at Iran during the war, prompting it to acquire and create a missile command. To this day, outlets continue to sing the praises of the IRGC [Iran's Revolutionary Guard Corps] officer who had the foresight to 'reverse engineer' Scuds, as opposed to firing them all at Iraq. These became the basis for Iran's missile forces."[20]

Conceivably, it is the Iran-Iraq War that defines and determines Iran's behavior regionally and globally. One argues how Iran influences the Iraqi government, defends Syria, and assists the Houthi rebels in Yemen. This behavior is the product of the eight-year conflict that shaped Iran's current security platform.

The fourth and final reason why the Iran-Iraq War still matters to Iran and the region is that the war reaffirmed Iran's 1979 Islamic revolution ideology. In fact, Ayatollah Khomeini wanted to ensure that subsequent generations learned the revolution's lessons and applied them when dealing with domestic and global issues that Iran might find itself engaged in. Khomeini's desire explains why the current Iranian leadership, who participated in the war, still refer to the conflict to inspire today's Iranian youth. In this way, the leadership instills a sense of duty and nationalistic pride and calls for citizens to embrace the principles of the revolution. The leaders also pursue a hidden agenda: to hold on to power. Could this explain why the thinking of the Iranian youth, who want to join the global community, represents a challenge, potentially even a threat, to Iran's religious establishment? I believe so. It seems that the survival of Iran's religious establishment, one that controls key sectors of the government, depends greatly on its ability to valorize the Iran-Iraq conflict. They must promote the argument that it was more than just a war, more than a mere exchange of missiles and firepower. They must affirm, rather, that it symbolized a divine test to promulgate the principles of the 1979 Islamic revolution so that those principles would live on for generations to come.

However, I argue against such thinking, especially now that economic sanctions have been lifted on Iran. Young Iranians are eager to join the world, share its vision and aspirations, and be part of the global citizenry, something Iran's hard-liners are nervous about and scared of. I address this theme in detail in chapter 5.

Significance

One should not underestimate the significance of the Iran-Iraq War. It offered a frightening glimpse of the dark shadows that would fall over the twenty-first century. While the West, mainly the United States, assumed that Iraq would decisively win the war, the West's hopes overall were shattered as Iran strongly resisted. In doing so, Iran reinforced the notion that when it comes to defending the homeland, Iranians of all stripes will put their differences aside and rally toward a common goal: fighting the enemy, whoever that might be.

Equally important, neither Iran nor Iraq was in a position to claim victory, military or otherwise, following the war. What the war provided, however, is political capital for both Saddam Hussein and Ayatollah Khomeini—more so the latter than the former. Indeed, the Iran-Iraq conflict paved the way for the revival of Shiism, especially after the 1979 Islamic revolution. The conflict cemented the rule of both leaders, Khomeini in Iran and Hussein in Iraq, thus shaping the Persian Gulf for years to come. One would think that with the removal of Saddam Hussein, the Shia revival would have prospered. To the contrary, it did not. Vali Nasr writes, "What Iran's revolution had failed to do, the Shia revival in post-Saddam Iraq was set to achieve. The challenge that the Shia revival poses to the Sunni Arab domination of the Middle East and to the Sunni conception of political identity and authority is not substantially different from the threat that Khomeini posed. Iran's revolution also sought to break the hegemonic control of the Sunni Arab establishment. The only difference is that last time around the Shias were the more radical and anti-American force, and now the reverse seems to be true."[21]

That said, what I find disturbing about the conflict is the violation of international law and the resulting killing as those violations were swept under the rug. Kurt Burch writes, "The eight-year war, the longest conventional war of the twentieth century, illustrated the horrific carnage of conflicts fought in defiance of international laws and norms concerning war. Human-wave attacks, missile attacks, and chemical weapons including lethal nerve gas, blister-agent mustard gas, and chemical fires killed one million soldiers on the battlefields and wounded many thousands of others. Hundreds of thousands of civilians suffered casualties, and millions became refugees."[22]

To conclude this section, I argue that since the end of the Iran-Iraq War, security analysts and academics have continued to struggle with the idea of how best to categorize the long-term impact of the conflict. No single or simple answer exists. The impact of the conflict is not limited to geopolitics but also encompasses economic, ideological, religious, and cultural dimensions. Of interest in this discussion is the political impact the conflict had mainly on the Gulf Cooperation Council countries given their proximity to both Iran and Iraq. Further, the ideological impact of the 1979 Islamic revolution should not be minimized. Monarchies in the region with minority Shiite populations became nervous that minority groups might rise up against the ruling families.

The concern in the region was even more prevalent when American foreign policy toward the Middle East collided with Iran's revolutionary objectives. It was as though two forces from opposite ends of the spectrum ran toward each other at breakneck speed and collided with full force. Nations and peoples beyond the borders of the Middle East felt the resultant aftershocks of the collision. Economically, the global energy sector also felt the effects as war raged around Iran's and Iraq's oil fields.

In the subsequent sections, I address in detail how the Iran-Iraq War contributed to US foreign-policy behavior toward not only Iran and Iraq but also the region at large. Navigating these dangerous waters has proved difficult for the US foreign-policy establishment. The invasion of Iraq, and Iran's nuclear program, serve as compelling examples of the dangers.

The Iran Nuclear Dossier

References abound when it comes to Iran's nuclear dossier. The term *nuclear dossier* does not refer to a set of documents or records; rather, I use the term throughout this chapter to refer to Iran's nuclear program. The following detailed narrative puts the Iran nuclear dossier within the security framework that came to define not only concerns in the West about Iran acquiring nuclear technology and becoming a nuclear state but also the global security landscape, mainly in the Middle East. Before delving deeper into this intriguing, divisive topic, let me say that I do not intend to address the historical framework of the nuclear dossier. Rather, I intend to focus this section's narrative on what Iran's nuclear dossier means from an international security perspective and international relations viewpoint. Further, I analyze where I see the program headed.

Iran's Nuclear Dossier and the International Community

I begin by highlighting some of the concerns and questions over Iran's suspected undisclosed experiments whose sole purpose is the development of nuclear weapons that were not addressed to the International Atomic Energy Agency's (IAEA) satisfaction. For instance, the IAEA wanted to know how Iran obtained P-1 gas–centrifuge technology, a critical component for the enrichment process. Iran provided unsatisfactory and vague explanations, and its disclosure to the IAEA that it received centrifuge designs from "intermediaries" in 1987 and 1994 was received with a heavy dose of skepticism. The IAEA's own investigation revealed that the delivery of centrifuge materials to Iran came from a clandestine supply network run by Abdul Qadeer Khan, a Pakistani nuclear physicist considered to be the father of Pakistan's nuclear program.[23] Similarly, the IAEA disclosed that, under the Joint Plan of Action, Iran failed to meet its obligations and honor its commitment to convert its enriched uranium from gas to powder.[24] In response, Ali Asghar Soltanieh, Iran's ambassador to the IAEA, described his country's nuclear dossier as "totally politically motivated."[25]

However, one must provide a brief account of how the P5+1 came to play a pivotal role in the negotiations with Iran over its nuclear program. It all started when three European countries (France, Germany, and Britain) proposed, in 2004 and 2005, to resolve Iran's nuclear program through peaceful means, including negotiations. Although negotiations during that time did not yield the desired result, Russia, China, and the United States decided to join the three European countries in 2006. At that juncture, those nations together formed what has become known as the P5+1—a reference to the five permanent members of the UN Security Council plus Germany.[26] Those moves set the stage for what became marathon negotiations between the P5+1 and Iran. The United States took the lead in those negotiations; however, that is not to suggest the United States negotiated unilaterally with Iran. Rather, it negotiated in consultation with the other P5+1 members.

This turn of events brought Iran's nuclear program to the forefront. Pundits and international security analysts with intimate knowledge of the potential impact a nuclear Iran could have on the geopolitical landscape in the Middle East have argued that eventually Iran will acquire the technical know-how to build a nuclear device. For instance, Steven Emerson, the executive director of the

Investigative Project on Terrorism, argues, "The dangers of an Iranian nuclear weapon are many. While the dangers of the conventional missile threat have been made clear, the danger of an Iranian bomb in the hands of terrorist organizations requires further analysis. The free world dismisses such threats at its own peril."[27]

Iran's response has always been focused on avoiding military confrontation with the West, mainly the United States. Iran also understood that the United States is unlikely to engage in another military conflict in the Middle East following its previous disastrous ventures, particularly its failure to achieve decisive victories in both Iraq and Afghanistan. The question is whether Iran understood early on that once its nuclear dossier was referred to the UN Security Council it would be presented as a security matter and perceived by the international community as such. That perception prevented Iran from pursuing alternatives suggested during negotiations. Some of the alternatives were based on whether Iran was willing to curb its uranium conversion, a key step in enrichment. It emerged in 2004 when the three European countries (France, Germany, and the United Kingdom) engaged in meaningful dialogue with Iran about its nuclear program. However, Iran did not hold to its end of the bargain and continued its clandestine nuclear activities despite disclosing in its proposal of January 17, 2005, that it would not pursue the development of weapons of mass destruction.

This change of perception compelled Iran to think in terms of what it needed to do to reach some sort of an agreement with the P5+1. As Iran's nuclear dossier became a priority for the IAEA, Tehran wanted to be perceived as negotiating from a position of strength rather than weakness and vulnerability. Iran's ambassador to the IAEA made this desire apparent, arguing, "The United States has the required power to inflict severe losses against us and to cause pain, but the Americans, too, are both vulnerable, and subject to suffering pain. All the same, if the Americans are determined to pursue that path, then let the ball roll."[28] In many Western capitals, especially in the United States, France, Germany, and Britain, the weeks that followed the referral of Iran's nuclear dossier to the United Nations Security Council in 2006 were marked by rigorous diplomacy concerning Iran's nuclear program.

It is imperative to briefly address this time frame of Iran's history, an era marked by a host of controversies. The international community perceived the election of Mahmoud Ahmadinejad in 2005 as a liability for Iran. Ahmadinejad was a controversial politician who contributed to the already negative image of Iran around the world. He had been widely criticized domestically for his poor management of the economy and globally for his incendiary statements about Israel. Further, the international community took notice of his disregard for human rights, evidenced by the crackdown on demonstrators in Tehran after the rigged election that resulted in his second term in office. To avoid drawing scrutiny over his failed economic policies and the rigged elections of 2009, Ahmadinejad blamed foreign countries, suggesting that they were behind the "green wave" revolution: a series of protests across Iran in the aftermath of Ahmadinejad's reelection in 2009. Demonstrators claim that the election results were fraudulent led to major protests in various Iranian cities. Internationally, Ahmadinejad exhibited hostility toward both Western and some Arab countries, most notably the United States, Great Britain, Saudi Arabia, and Israel. His hostility toward the Jewish state was no more evident than when he declared that the Holocaust had never occurred.

According to Farideh Farhi, an independent scholar and affiliate graduate faculty member at the University of Hawaii at Manoa, Ahmadinejad's government manipulated the June 2009 election results. Manipulation of the elections aimed to (a) boost Iran's leverage during its nuclear negotiations with the P5+1 over its nuclear program, and to (b) demonstrate to the rest of the world that Ahmadinejad's government was not yielding to the West's pressure.[29] While Ahmadinejad might have been popular inside Iran among hard-liners, his international reputation was negative, thus contributing to the perception that Iran was a pariah state.

Conversely, Ahmadinejad's thinking was that if he played hardball with the West and demonstrated his toughness through his rhetoric and empty promises, it could cement his legacy and pave the way for yet another hard-line successor. Another point that merits emphasis is that Ahmadinejad's aggressive, ill-conceived policies, extreme stance, and pretentious views were no match to the supreme leader, Ayatollah Khamenei. However, the latter expressed his support for Ahmadinejad mainly during Friday prayer sermons to avoid any public discord in the ranks. Farideh Farhi argues, "It is often said in Iran that whoever makes a deal with powerful outside players, above all Washington, to end the Islamic republic's international isolation will tighten his grip on the state for good. Ayatollah Khamenei has been less bombastic than Ahmadinejad in blaming foreign actors for the 'green wave' that engulfed Tehran and other Iranian cities in protest of Ahmadinejad's anointment as president in June. But Khamenei was explicit in his Friday prayer speech on June 19 that, in general, he stands with the cocky Ahmadinejad and not more conciliatory conservatives in matters of foreign policy."[30]

On the economic front, Ahmadinejad's economic policies were disastrous and had a negative impact on the country's long-term economic strategy. Some of his campaign promises advocated sharing Iran's oil wealth more evenly with all segments of society. Simultaneously, he harshly criticized, even condemned, capitalism as an immoral system. To the surprise of many observers, and in an about-face, in 2010, Iran embarked on major economic reforms, sweeping measures that had not been implemented since the beginning of the 1979 Islamic revolution. These measures consisted of providing government subsidies in energy and consumable goods (oil, bread, and so on) to poor Iranian voters to keep up with the changes taking place in the global economy. These economic reforms had the support and blessing of the supreme leader, Ali Khamenei, who referred to 2011 as a "year of economic jihad."[31]

On the nuclear front, Ahmadinejad vehemently advocated for Iran's right to develop nuclear technology. On many occasions, he expressed his support for his country's nuclear program. While Ahmadinejad had manifested his aversion to the West on many occasions, his distrust in the IAEA was evident when he issued an order disallowing IAEA inspectors from freely visiting Iran's nuclear facilities. That came on the heels of Iran's long-range missile test of May 2009, with Ahmadinejad claiming, "Iran is running the show."[32]

With Ahmadinejad in office, many in the West did not expect changes in his stance regarding Iran's nuclear program. In a sudden change in policy, the United States, France, and Russia proposed a UN draft resolution in October 2009 that would move Iran and the West to a comprise that addressed both parties' interests and concerns. Iran needed a nuclear reactor for civilian use. However, the West worried that Iran's nuclear program could quickly become a military venture if not contained or eliminated.

Observers of global affairs often questioned how the Obama administration, especially in its first year in the White House, would handle the Iran nuclear dossier. Would President Obama pursue the same path his predecessor, Bush, had adopted, or would Obama embark on a new strategy? At that time, both the Iranian and US administrations were in transition. Ahmadinejad was reelected under questionable circumstances, while President Obama was elected to his first term in office. In 2009, Abbas Milani wrote, "What lies ahead for the U.S.–Iran relationship as both countries potentially transition in two new administrations? Even if Ahmadinejad is reelected, his badly tarnished image, and the now evident vapidity of his economic populism will make him less giddy with the arrogant self-righteous bravura of his first term, and arrogant leaders filled with self-righteous piety make bad diplomatic interlocutors. What sort of policy should the Obama administration adopt toward Iran?"[33] Despite all that, in a surprise change in policy, Ahmadinejad agreed to a deal in October 2009: "We welcome fuel exchange, nuclear co-operation, building of power plants and reactors and we are ready to cooperate."[34]

That change in policy paved the way for Ahmadinejad's successor, Hassan Rouhani, a well-seasoned politician and an insider (a former diplomat, Islamic cleric, academic, and a lawyer), to fulfill his campaign promise to end hostilities toward the West and reach an agreement with the P5+1 over Iran's nuclear program, which he did in 2015. Before his election, Hassan Rouhani promised to unshackle the country from the economic chains of sanctions that had crippled it for decades, since the revolution of 1979. Before the election of Rouhani, Ahmadinejad's administration had failed to achieve any progress regarding Iran's negotiations with the West. This lack of progress stemmed from Ahmadinejad's unwillingness to seek a compromise with the West. His hard-line stance was apparent in his attitude and approach to negotiating with the P5+1 members, which left little room for compromise. While there were tensions between Ahmadinejad and the supreme leader, Ali Khamenei—along with Hashemi Rafsanjani, who was president of Iran from 1989 to 1997—they managed to keep these tensions from becoming public knowledge. During most of the phases of nuclear negotiations with the West, Ahmadinejad exhibited his willingness to disobey and subtly challenge the supreme leader's orders. Hanif Zarrabi-Kashani argues, "There was always a chance that Ahmadinejad would not obey (Khamenei). Rafsanjani or (Akbar) Nateq-Nouri might have a difference of opinion; but they always follow the command of the Supreme Leader."[35] During the negotiations between Iran and the P5+1, Iran attempted to portray itself as a tough negotiator that would only be satisfied with a compromise to its own benefit. This approach backfired when Iran refused to yield to the demands from the UN Security Council to halt enrichment, leading to additional sanctions. Ahmadinejad's approach to the negotiations further hindered Iran's ability to reach a compromise with the West over its nuclear program. Further, it made the task of undoing the damage and regaining some level of trust from the West in order to negotiate in good faith even more complicated for the next administration, that of Hassan Rouhani. Another example that demonstrates how Ahmadinejad was set, from the start, on a course not to find a solution to the impasse between Iran and the West was how he tried to inject issues into the negotiations that were irrelevant to the issue at hand: Iran's nuclear program. For instance, Ahmadinejad requested that members of the P5+1 state on the record their position vis-à-vis Israel's nuclear arsenal—a request that was irrelevant to what the West and Iran were trying to achieve through negotiations.[36] Evidently, Ahmadinejad failed to remember that, for a long time, Israel adhered to a policy of nuclear opacity—a policy that neither confirms nor denies the existence of a substantial nuclear arsenal.

Despite the vicissitudes of the marathon negotiations, the election in 2013 of Hassan Rouhani, who was intimately involved in the negotiations, having been part of previous administrations, was seen in some circles within Iran as a step toward resolving the nuclear dossier with the West once and for all. His moderate tone and promise to end hostilities toward the West helped his case. Clovis Maksoud argues, "While some of the major issues and policies remain in the realm of Ayatollah Ali Khamenei's authority, the overall outcome of the elections provides an opportunity to improve relations with the international community. Although this can be considered for the moment a tentative impression, Rouhani's election has to be generally considered a positive development. Of course, while the welcome of many western countries will be tepid, the chance for future improvement is now within the realm of possibility."[37] The consensus at that time was that reaching an agreement with the West over Iran's nuclear program was like clinging to a mirage.

Iran's Nuclear Dossier and the Future of the Nonproliferation Treaty

While Iran's nuclear dossier took on a life of its own, global-affairs strategists, international security analysts, pundits, and academics expressed concerns, publicly and privately, over whether Iran's nuclear program, if left unresolved, could affect the future of the nuclear nonproliferation treaty (NPT). Many in the West wondered whether Iran, a signatory member, would abide by the regulations set forth in the NPT or pursue a path similar to that of North Korea, Pakistan, and India.

Since Iran's nuclear activities were reported to the UN Security Council (UNSC), academics and international security analysts have asked whether the case of Iran would be fairly approached, judged, and dealt with in an international system marked by inconsistencies. The other argument has to do with questions of legitimacy. Specifically, since Iran is a signatory member of the NPT, is it not within its rights to pursue civilian nuclear technology? The international system's approach toward Pakistan and India when they tested nuclear weapons exposed the flaws inherent within the language of the NPT. Of interest in this discussion is the possible outcome if the West and Iran had failed to reach an agreement. On both sides of the Atlantic, security analysts argue that there is no basis on which they could predict the outcome of Iran's nuclear dossier. Others, including Nasser Saghafi-Ameri, a veteran Iranian foreign policy and international security analyst and a guest contributor for Middle East Institute in Washington, DC, suggest the possibility of three scenarios regarding Iran's nuclear dossier: "As things stand today, three scenarios could be envisaged regarding Iran's nuclear dossier. Each of those scenarios obviously will have different outcomes, not only for Iran and other concerned parties but also especially for the NPT. To simplify the issue, we shall consider Iran's nuclear dossier under three models of behavior, namely, the North Korean model, the Libyan model, and the Japanese model."[38] Before going further, one must briefly shed light on the three models Iran might have pursued in the event negotiations with the P5+1 failed.

The first model is that of the Democratic People's Republic of Korea (DPRK or North Korea). Exercising its right to withdraw from the NPT on March 12, 1993, North Korea embarked on a policy of defiance by declaring its intentions to enrich plutonium at a military-grade level and to produce a nuclear bomb. What is unclear, however, is whether Western intelligence agencies can determine how many nuclear bombs the DPRK possesses. This ignorance is attributed to two main factors. First, the international community cannot infiltrate the DPRK to get firsthand knowledge of what the DPRK has. Second, the DPRK leadership's tight control over the population prevents any contact with the outside world that may illuminate what sorts of activities the DPRK is engaged in. In fact, the DPRK withdrew from the NPT, rendering any means of control unavailable. Articles within the treaty may have contributed to the DPRK's justification for withdrawal. George Bunn and John Rhinelander argue, "Article X of the NPT provides a 'right' to withdraw from the treaty if the withdrawing party 'decides that extraordinary events, related to the subject matter of this [t]reaty, have jeopardized the supreme interests of its country.' It also requires that a withdrawing state-party give three months' notice."[39]

International security analysts continue to argue that DPRK's long-term objective is to complete its nuclear program. Simply stated, the previous negotiations with the six-party forum, which includes China, Japan, North Korea, South Korea, Russia, and the United States, were intended to mask DPRK's nuclear ambitions. The military leadership in North Korea wants to ensure that the United States would never entertain a military attack on the DPRK. It seems likely that the West will always have reservations about engaging North Korea militarily as long as the West cannot confirm the DPRK's nuclear capabilities. One must not forget that the DPRK will not embark on any policy without consulting with its patron, China. It is unclear to elites why, if the DPRK engages in back-and-forth diplomacy with neighboring South Korea, it has taken such a bizarre diplomatic approach. Alexandre Mansourov offers, "I assume that the North's motivations in dealing with the South are primarily strategic, with military and domestic security factors being secondary, and economic considerations being tertiary in significance."[40]

Some argue that North Korea never trusted the IAEA, viewing it as one of the United States' tools for executing the West's—mainly the United States'— hostile policy toward North Korea. This was evident when on February 9, 2005, the DPRK announced its possession of nuclear weapons. It was suggested that, after North Korea released the news, the United States would shelve some of its military plans against North Korea lest US military forces or interests in the region come under attack.

I am unconvinced that Iran would have followed the North Korea model if negotiations with the P5+1 had failed. I argue that the failure of the negotiations was not an option for not only Iran but also China and Russia, the main beneficiaries of Iran's nuclear agreement. For the sake of argument, let us assume the negotiations had not produced the desired result—a complete halt of Iran's nuclear activities. Iran would have had no choice but to withdraw from the NPT and go it alone and deal with any consequences.

Similarly, I argue that if Iran had pursued the North Korea model, that would not have hindered Iran's ability to continue clandestinely purchasing and developing advanced nuclear and missile technologies to enhance its military capabilities. As a matter of fact, before the nuclear agreement between Iran and the P5+1 in the summer of 2015, Iran engaged in clandestine activities to illicitly procure materials necessary for developing its nuclear and ballistic missile programs. The question remains why, when it was revealed that Iran had indeed engaged in illicit activities during the negotiation phase, the West chose not to react. Did the West want to avoid embarrassment? Or was it yet another intelligence failure? Unfortunately, we might never know the answers to these questions. Alas, Iran's violations demonstrate not only the West's lack of political will but also how divided the negotiating countries (P5+1) are. Benjamin Weinthal and Emanuele Ottolenghi suggest that more evidence of Iranian violations has now surfaced. Two reports regarding Iran's attempts to illicitly and clandestinely procure technology for its nuclear and ballistic missile programs have recently been published. They show that Iran's procurement continues apace, if not faster than before the Joint Plan of Action was signed in November 2013. But fear of potentially embarrassing negotiators and derailing negotiations has made some states reluctant to report Tehran's illegal efforts. If these countries hesitated to expose Iran during the negotiations, it is likely they will refrain from reporting after a deal is struck.[41]

Conversely, observers are scratching their heads as to why the final draft of the Iran nuclear agreement with the West did not specifically address the status of the Parchin military facility—the compound where most of Iran's past nuclear military activities were carried out. Once again, we are left to wonder why the West failed to bring this issue to the forefront. Was it the ineptitude of the negotiators? Or was it simply a way to score political points? Whatever the case may be, it is evident to those who read the final draft, including Thomas Moore, an arms control specialist and former Senate Foreign Relations Committee staff member, that the IAEA refrained from including in the draft language a reference to Iran's past nuclear activities as it relates to military dimensions. "The IAEA's resolution of the possible military dimensions of Iran's nuclear program should precede the deal, not by months but by as much time as it takes to verify the absence of Iran's [past military work], including the full historical picture of its program,"[42] Moore argues. Others, such as Fred Fleitz, a former CIA analyst, State Department arms control official, and House Intelligence Committee staff member, made it clear that the language of the draft, as it pertains to the lifting of sanctions, will allow Iran to pursue rearmament.

The second model is that of Libya. Security analysts argue that Libya's decision to give up its quest for nuclear armament was in the works during Bill Clinton's second term in office. It is a false assumption that President George W. Bush's policy to spread democracy in the Middle East and invade Iraq made it possible for Libya to abandon its nuclear program. Historians will have to judge whether there was ever a link between the invasion of Iraq and Libya relinquishing its nuclear aspirations. Scholars including Flynt Leverett of the Brookings Institute, an American think tank, who worked for the Bush administration until early 2001, argues, "By linking shifts in Libya's behavior to the Iraq war, the president misrepresents the real lesson of the Libyan case."[43]

While a few pundits try to convince themselves that Iran would have followed the Libyan model, conventional wisdom suggests that the chances were slim to none. The reason is that mistrust toward the United States prevails in the minds of Iranians of all stripes. Iranian mistrust of the United States traces its roots to the United States' orchestration of a coup against Mohammad Mossadegh in 1953. Americans tend to forget that our foreign-policy behavior sometimes comes back to haunt us decades later. Such outcomes demonstrate a lack of strategic vision and forward thinking and the willingness to achieve short-term gains instead of long-term objectives. The invasion of Iraq makes this point. Efforts to build trust between Libya and the United States preceded any talks about nuclear disarmament. The United States also removed Libya from the list of states sponsoring terrorism as a reward for abandoning its nuclear program.

Another reason Iran would not have followed the Libyan model is that the two countries differ in terms of regional status, economic power, religious ideology, and military capabilities. In addition, Iran knew that if China and Russia got on board and supported it during the negotiations, there would be major military, economic, and diplomatic benefits. That is exactly what is happening now between Iran and both China and Russia. Iran has since signed a total of seventeen bilateral agreements with the two countries, mainly in the energy and military sectors. US policy makers were misguided in their assumption that they could sway or manipulate the Iranians as easily as they may have done with the Libyans, the Iraqis, the Egyptians, and the Saudis. Alas, the case of Iran proved to be far more complex than the Washington establishment anticipated.

The third model is that of Japan. Many argued that the Japan model was the one Iran would be likeliest to pursue; however, there was no inclination in Tehran toward that end, at least during the phases of the nuclear negotiations. Like other security analysts, I argue that the Japan model proved that signatory countries in the NPT could indeed balance the application of nuclear civilian technology with that of the requirements for nonproliferation. Simply stated, a NPT signatory country could pursue civilian nuclear technology as long as it guaranteed that there would never be a switch from civilian technology to military application. For this reason, I believe that had the negotiations between Iran and the P5+1 members over the former's nuclear program failed, Iran would have likely pursued this model.

The Japan model displays the efficiency and success of the NPT's safeguard measures. Would the Japan model be suitable for Iran if it had pursued this path? I doubt it. There is, however, a consensus among nuclear experts and security analysts that the Japan model constitutes the basis for "nonproliferation culture." However, the desire within Iran concerning its nuclear program was to reach an agreement, have the economic sanctions lifted, and reintegrate into the world economy. We should remember that the Japan model is based on that country's own horrific experience with the might and power of nuclear weapons. The

United States' dropping nuclear bombs on Nagasaki and Hiroshima scarred Japan forever. Further, it made Japanese political elites pragmatic about acquiring weapons of mass destruction, be they nuclear or hydrogen bombs.

Against this backdrop, I argue that, among the three models presented herein, the Japan model would have been the likeliest possibility had Iran decided to go it alone. However, against all odds, an agreement between Iran and the P5+1 was reached in the summer of 2015. The question now is what is next for Iran now that the sanctions have officially been lifted and floods of investments are in the works, paving the way for Iran's reintegration into the world economy.

Make no mistake: the deal Iran reached with the West over its nuclear program has not been well received among hard-liners; yet most Iranians welcome this deal because they strongly believe that the lifting of sanctions will eventually raise the standard of living for most Iranians, an outlook the hard-liners do not share. Yes, economic opportunities await Iran, but they will not come too fast. Iran's economy consists of a series of knots that will take time to loosen. However, since the government, with all its entities and institutions, is firmly in the hands of the religious establishment, tensions are already flaring up given how access to over $100 billion in once-frozen assets might not trickle down to average citizens. Rather the access to these funds will benefit the elites, the religious establishment, and the military apparatus, especially Iran's Revolutionary Guard Corps.

I believe that the chants of "Death to America" and "Death to Israel" are for domestic consumption. However, the fact that Supreme Leader Ali Khamenei has used these phrases in the religious sermon he delivered following the deal highlights his anxiety that Iran's nuclear deal with the West could, indeed, open up the country to the outside world. Khamenei is well aware of how Iranians, especially the youth, are slowly but surely embracing the outside world and moving toward assimilation. That outcome terrifies Iran's hard-liners, including the supreme leader. The reason is that the survival of the theocracy in Iran depends on convincing the masses that any opening to the outside world goes against religious teachings (which is untrue based on my deep understanding of the Islamic culture) and will negatively impact and corrupt society morally, socially, and ethically.

It's déjà vu all over again, as the saying goes. Almost forty years ago, religious hard-liners had to find an ideological reason to justify their stance: ongoing aggressive posturing against Israel, the United States, and its allies in the region. The ideological argument manifested through the teachings of the Islamic Revolution was then—and is now—employed in order to ensure the hard-liners a beneficial outcome. Fast-forward: the current supreme leader, Ali Khamenei, finds himself dealing with an outcome similar to the one that made possible the Islamic revolution of 1979. He is trying to demonstrate to the religious establishment—and Iranian society, for that matter—that he wants to preserve and ensure both Iran's national interests and identity after the nuclear deal. "In a speech at a Tehran mosque punctuated by chants of 'Death to America' and 'Death to Israel,' Khamenei said he wanted politicians to examine the agreement to ensure national interests were preserved, as Iran would not allow the disruption of its revolutionary principles or defensive abilities," argue Babak Dehghanpisheh and Bozorgmehr Sharafedin.[44]

However, Khamenei's bravado, absurd prating, and rhetoric are far from reflecting reality. Many in Iran, especially the youth, are growing disenchanted with a regime that limits their potential to explore the world, improve their living conditions, acquire new skills, and forge new paths in science and technology. I believe that Iran's religious establishment assumed that a nuclear agreement with the West would not be reached. When it was, it changed the calculations and the narrative within the establishment. The religious establishment is now concerned with how to deal with the aftermath of the agreement, with Iranian society's growing support for the deal, and with the possibility that many Iranians are on the verge of seeing that the end is near for the religious establishment.

But not so fast. One should remember that key government institutions, namely the intelligence apparatus, the military, and the economy, are under the control of the supreme leader. Anyone who thinks that Ali Khamenei will give up his power for the sake of global assimilation, integration, and democracy—in whatever form or shape it may take—is mistaken. Could this explain why, during Friday sermons, the slogans "Death to America" and "Death to Israel" are frequently repeated? It is a reminder, a psychological conditioning for the society in the hopes of keeping the roots of the Islamic revolution alive. However, I strongly believe that it is only for domestic consumption because the supreme leader and his entourage understand that any attacks on US interests in the Middle East—or Israel, for that matter—would amount to suicide. Simultaneously, in the United States, we have demonstrated in the past few years that we talk more than we act; we are great at using threatening language but very weak at following through. Basically, our threats amount to empty rhetoric. The result is a loss of credibility. The Obama administration backing down from its "red line" approach to Syria's use of chemical weapons evidences the kind of weak inaction that results in a loss of credibility.

China and the Potential for Economic Expansion in the Middle East

While the United States argued on behalf of the global powers for a comprehensive agreement with Iran over its nuclear program, Russia, a member of the P5+1, knew all along that it could stall US efforts in reaching an agreement with Iran. Further, Russia realized that it is now or never to reassert itself in the region with countries that were once under the Soviet flag. In these ever-changing dynamics, Iran's nuclear dossier has triggered the intervention of members of the

European Union, such as France and Spain, besides other major powers, mainly China and Russia.

For instance, China's ambassador to Iran expressed his country's support for Iran's right to nuclear technology. The Chinese ambassador to Tehran, Leo Jen Tung, argued that, as a permanent member of the United Nations Security Council, his country has always supported Iran's right to use nuclear energy for peaceful purposes.[45] China's declaration of support highlights two main factors: (1) China is making its voice heard through the influence of global affairs, thus raising its global diplomatic profile and sending a message to the West that it must take China's interests into consideration; (2) by stating its support for Iran in the United Nations Security Council, China is sending a strong message to the world, mainly the West, that it cannot ignore the role Beijing plays in international forums. That move highlights China's economic, security, and political interests in the region. After the nuclear agreement between Iran and the West, China rushed into signing seventeen economic treaties worth about $600 billion.[46] As argued in my previous writings, it is just a matter of time before China enters into a military treaty with Iran, one that would not only provide China with military bases in the region from where to project its strength but also challenge US hegemony in the region.[47]

China realized early on that Iran's nuclear dossier provides it a great opportunity to pursue its strategic interests in the wake of its economic rise and military expansion in the South China Sea. The question is whether China considers using Iran's nuclear dossier as a platform on which to launch its strategic objective in the greater Middle East, that may compel the Washington establishment to return to reconsider its foreign-policy options. Current dynamics on the ground suggest that China's strategy is on target. Chinese president Xi Jinping's official visit to Tehran in January 2016 and the half-trillion dollars in bilateral trade over the next decade demonstrate China's strategy at work.

China's increasing footprints in the Middle East reflect its increasing presence on the global stage and its sense of power. Chinese president Xi Jinping's statement of September 23, 2014, in which he issued an order that military leadership be ready to win a war, while ambiguous, is worrisome. I am not surprised by Jinping's statement. Such statements do not happen in a vacuum. China's economic preeminence heralds an inevitable expansion of its military power. Modernization of its military is part of China's strategic plan. It has already established closer ties with India, Pakistan, and Russia. Its cooperation with the latter is based primarily on China's desire to redirect US focus away from Asia, which it has succeeded in doing thus far.

Like other military experts and security analysts, I anticipate that China's military technology will continue to grow in the years ahead. China has recently upgraded its nuclear stockpiles by introducing multiple warheads. It now has second-strike capability. This comes as the United States tries to deduce what Jinping's announcement means. Is China preparing for a major war? To better understand these dynamics, one should study China's tactical and strategic moves. For instance, China has embarked on a major development of infrastructure in the South China Sea, including building an airstrip big enough for military aircraft. I argue that China's buildup allows Beijing to demonstrate its strengths, a move that has already introduced fear and anxiety in neighboring countries—Japan, Brunei, the Philippines, Taiwan, Malaysia, and Vietnam. China is sending a message to the United States that its days of influence over the Pacific are numbered. That China's buildup in the South China Sea is not limited to basic infrastructure (roads, water pipes, and so forth) but also includes military structures (airstrip for military aircrafts, radar towers, and so forth) undoubtedly communicates to the West China's territorial ambitions and desire to control waterways. As I argued in previous writings, China's move is no surprise. Its economic expansion is translated into a military enlargement, a move supported by increased defense spending that will double by 2020.

Furthermore, I must place these geopolitical changes in the Pacific within the context of the impact that China's nuclear upgrade might have on the global balance of power. Based on my research, I predict that the size of the Chinese navy will surpass that of the United States by 2020. Indeed, China plans to produce new warplanes, destroyers, and assault vessels. Of particular worry is China's ability to acquire a sea-based nuclear capability, including missiles that have a strike range of some 4,567 miles.[48] Of great concern is that, should a military conflict between the United States and China erupt over the South China Sea fiasco, which I do not foresee, China's upgraded missiles can reach anywhere in the United States if launched from the coastal waters off Hawaii. The US government therefore should avoid, at any cost, unnecessarily provoking China, especially when it comes to Taiwan. The United States has already acted ill-advisedly with Russia over its annexation of Crimea, and it should not make more enemies than it can afford.

Could China's rise upset the global balance of power? Some of its neighboring countries are already jittery about the possibility of a military confrontation between China and whoever challenges it over its sovereignty over the South China Sea. While the dispute over the ownership of the islands is not new, the current tensions highlight the possibility of a direct confrontation among superpowers, a scenario that would have devastating consequences.[49] The United States should have a greater understanding of why China would go to war over the islands. In addition to China's complex history in the region and the economic importance of the islands, China's threats reflect its main strategic objective: reducing US influence in the Pacific.

The importance of the islands in the South China Sea should not be underestimated. The Pacific passageway is used to transport half the world's shipments of goods and one-third of its oil and gas. As a result, China realizes the importance of having full control over those waters and thereby imposing its will by controlling access routes to markets, be they in Europe and the Middle East or South Asia and the Pacific. These dynamics make me wonder whether the United States is willing to take a potentially costly military risk, as neither the United States nor China nor any other major power knows what the future may hold for this "boisterous" region and the brewing international rivalry in the waters of the Pacific.

Moreover, China's shift in posture is calculated and reflects two main points. First, China's expansion and military buildup in the South China Sea is, in my opinion, an extension of its economic growth. This economic progress contributes to a shift in wealth from the West to the East. Equally important, China intended to take full advantage of these changes in the economic landscape to circumvent and replace the US dollar with its currency, the yuan, in global oil transactions, a strategy it succeeded in implementing. Do you recall when China hosted, in June 29, 2015, the signing ceremony of the fifty-seven founding member countries of the China-led Asian Infrastructure Investment Bank? The establishment of this financial institution saw China contribute nearly $30 billion of the institution's $100 billion capital base. This move gave China a say, allowing Beijing to have 25 to 30 percent of the bank's total votes. Surprisingly, Australia and the United Kingdom decided to join the new bank, a move that highlights the widening gap among countries in the West. Katherine Murphy writes, "When Britain joined the AIIB in mid-March, the White House issued a rare public rebuke, declaring its expectation that the UK would use its influence to ensure that high standards of governance are upheld. The US sees the AIIB as a vehicle for China to exert influence in the region."[50]

Second, like most major powers before it, China's economic rise will inevitably translate into its military expansion, as evidenced recently. China's vision, strategy, and forward thinking, in my opinion, have contributed to this shift. Make no mistake: China's rise did not happen in a vacuum. China's long-term vision and its ability to create and rigidly pursue economic, military, and financial strategies over the past few decades have led to this strength. In support of my latter assertion, recall that China conducted a successful test of its hypersonic nuclear Wu-14 missile (more than 7,600 miles per hour), a test that rattled and confused Washington.[51] Equally important, other countries have embarked on acquiring similar technology that would undoubtedly pose a threat to the security of the United States. According to *Jane's Intelligence Review*, Russia's top-secret program Project 4022 involves the development of a nuclear-capable hypersonic missile. If deployed, the missile would be able to reach and destroy large targets within minutes.[52] I am certain the United States is working on, if it has not already developed, similar, perhaps far superior, capabilities.

The takeaway is that China's assertive stance regarding the South China Sea suggests that the United States and China will eventually collide if their interests are threatened. What concerns me the most is that China's already assertive security posture, if it persists, signifies a US leadership decline in the Pacific as yet other crises loom on the horizon. Despite all that, China strikes me as a pragmatic nation, one that takes short but assured steps rather than long but unsteady strides. Moreover, when it comes to its security interests, be they in the Middle East or South China Sea, Beijing will likely be disciplined in how it proceeds to gauge its military interventions, whatever the causes may be.[53] The American foreign-policy establishment should be careful not to antagonize China because the United States could discover that a military conflict with China is no small matter.

Russia's Double-Edged Political Game

All along, Russia has calculated that, given that it wields veto power in the United Nations Security Council, it can influence the outcome of the nuclear negotiations between Iran and the P5+1 toward achieving its geopolitical and economic aspirations. For instance, Russia ignored calls by the United States not to deliver an advanced air-defense system to Iran. Instead, Russia unexpectedly replaced the delivery of its S-300 air-defense system to Iran with that of the Antey-2500, a more sophisticated, more powerful defense system.[54]

I also argue that the Kremlin fully understands that it is now or never for Russia to reassert itself on the global stage through the ongoing turmoil in the Middle East. The civil war in Syria and the agreement of the West with Iran over its nuclear program provide Russia the much-awaited opportunity to secure political influence backed by a strong military presence. To convince critics who argue otherwise, one could evaluate the military hardware Russia brought to Syria: Yakhont, a 6.7-meter-long (22-foot) missile with a range of 290 kilometers (180 miles), carrying a high-explosive or armor-piercing warhead, and warplanes (thirty-four of them) that match or exceed the performance of our F22s and F18s.[55] Against this backdrop, the Washington establishment, military and civilian alike, must neither underestimate nor minimize the role Russia continues to play in the fast-changing geopolitical landscape of the Middle East.

Russia employed similar tactics at the dawn of World War II. Do you recall the Molotov–Ribbentrop Pact, named after Soviet foreign minister Vyacheslav Molotov and German foreign minister Joachim von Ribbentrop? The nonaggression treaty allowed Josef Stalin's Soviet Union and Hitler's Nazi Germany to divide Poland. Could a similar scenario await Syria? Possibly, though, at least this time around, there are more than two key players.

Could this explain why Russia has always insisted on pursuing negotiations with Iran rather than applying military force to prevent Iran from acquiring a nuclear device? Deputy foreign minister Sergei Kislyak argues, "Russia still believes that Iran's nuclear dossier should be sorted out by means of negotiations."[56] When compared to China's motives, Russia's motives provide far greater insight into a nation that wants to use Iran's nuclear program to advance its own agenda. Consider, for example, the civil war in Syria, where Russia and Iran work together to ensure that Bashar al-Assad stays in power. This is evidenced by how Russia and Iran collectively provide military, financial, diplomatic, and logistical support to ensure that the al-Assad regime prevails in its war against Western-supported rebels. Retaining the al-Assad regime allows both Iran and Russia to pursue similar interests but different objectives. To illustrate how far Russia is willing to go to advance its geopolitical interests, it vetoed many United Nations Security Council resolutions calling for the intervention of an international force in the Syrian theater.

Similarly, Russia's long-term strategy has never been about just reaching an agreement over Iran's nuclear program; rather, it wants to ensure its seat at the table and its say in other global conflicts, including the civil war in Syria and upheavals in Yemen, among others. Unequivocally, I argue that Russia sees in a chaotic Middle East a great opportunity to regain lost influence and reestablish ties with countries in the region once under its umbrella. I also argue that, from its perspective, Russia uses the success of the negotiations, which led to an agreement between the West and Iran over its nuclear program, as a platform on which to strengthen Russia's argument for pursuing a similar strategy in dealing with other contentious global conflicts. That was evident when the Russian permanent representative to the UNSC, Vitaly Churkin, said, "We would like this invaluable experience of joint actions, which is not burdened with geopolitical calculations, to be used for resolving other crisis situations in which success can only be reached through cooperation. Russia is ready for that."[57]

The counterargument is that Iran's nuclear dossier reveals the extent of the divisions that exist between the major powers: on one side are the United States, France, Germany, and Great Britain, and on the other are Russia and China. Yet, the rest of the world, except those with firsthand knowledge of the details of the negotiations, has no idea of the internal tensions among the P5+1 members. For instance, few nations are aware that Russia and China have vehemently refused to support added sanctions on Iran. That suggests that Russia and China want to ensure that an agreement between the West and Iran is reached so they can benefit economically and politically. As of this writing, China has signed seventeen economic treaties with Iran worth about half a trillion dollars over the next decade. I see no reason for Iran, now that sanctions are officially lifted, not to award major contracts (energy and others) to those two countries knowing that they provided it political cover in international forums during the marathon nuclear negotiations.

Russia and China strengthening their economic ties with Iran comes at the detriment of the United States' strategic interest in the region. In fact, the expansion of Russia and China's economic and military ties in the Middle East has baffled the Washington establishment and driven pundits to question what it all means. The thinking has always been that the United States' influence in the region is unwavering, its leadership unshakable, and its ability to influence its allies unquestionable. However, reality suggests otherwise. Current dynamics not only in the Middle East but also around the world suggest otherwise. For instance, in the Middle East, the United States failed to exhibit resolve and leadership when the Obama administration was unwilling to punish the al-Assad regime when it used chemical weapons on Syrian people in December 2012. The United States' inaction not only proved a major strategic mistake but also demonstrated endless flexibility and a lack of American resolve.[58]

The challenge to US global leadership and loss of credibility is not limited to the Middle East but rather extends to other corners of the globe. For instance, in an unexpected move, two of the United States' closest allies, Great Britain and Australia, joined the Asian Infrastructure Investment Bank. Many observers argue that that move signals the two countries' endorsement of China's efforts to challenge the United States' most dominant financial institutions: the World Bank and International Monetary Fund. Already there is talk in some Western capitals that the United States will not be the dominant Pacific military power indefinitely, nor the world's foremost economic powerhouse.

Given these force-producing events, we are witnessing a different Middle East, a new geopolitical landscape that puts American foreign policy's ambiguity toward the region at the forefront of the global debate.[59] These developments did not occur in a power and security vacuum. Many conflicts unfolded throughout the region: ongoing sectarian violence in Iraq, security and political instability in Egypt, a failed state in Libya, raging civil war in Syria and Yemen, and, more recently, strife and a cut in diplomatic relations between Iran and the Kingdom of Saudi Arabia.

Iran's strategic concerns stem from two main factors: (1) the US invasion of Iraq and the indecisive war in Afghanistan; and (2) the establishment of American semipermanent military bases in addition to Iran's entire coastline being strongly and effectively dominated by the presence of US Fifth Fleet.[60]

The clash of perception and mistrust in both Washington and Tehran put both countries at odds when it comes to cooperation, for instance, in defeating ISIS, finding a solution to the ongoing Syrian civil war, resolving the ongoing sectarian violence in Iraq, and preventing further deterioration of conditions in Yemen. I believe Iran will play a pivotal role in Middle Eastern affairs in the coming years. It is up to the United States to figure out how to reduce this clash of perception in order to move forward. Alas, if both counties cannot reach some understanding, other countries, such as Russia and China, might encourage Iran and the United States to seek a more realistic view concerning security and strategic interests. Should this scenario become a reality, it will boost Russia and China's diplomatic egos on the world stage, thus compelling them to address other global conflicts that they would have refrained from addressing otherwise.

Current events in the Middle East highlight the United States' inability to manage the ongoing turmoil in a region marked by constantly shifting loyalties and endless chaos. Iran's nuclear dossier had been in the works before the conflicts mentioned above occurred. Nonetheless, Iran's political elites saw an opportunity through the nation's nuclear program to take advantage of the turmoil the region continues to witness.

Four Points that Help Iran's Position on Its Nuclear Dossier

It is fair to say that Iran's ability to foresee and influence events on the ground as they occur has helped it throughout the phases of the negotiations over its nuclear program. This can be attributed to four main points.

First, the US invasion of Iraq and the disastrous outcome it produced allowed Iran to play on multiple fronts. The invasion got rid of Iran's worst enemy, Saddam Hussein, and allowed it to drag out negotiations over its nuclear program, knowing how entangled the United States was in Iraq. Furthermore, ongoing sectarian violence in Iraq helped call into question the United States' leadership and strategic vision. For instance, before the invasion of Iraq, did the United States think of what it would mean for the Middle East if Shia ascended to power in Iraq? Did the United States consider what would happen if negotiations with Iran over its nuclear program failed? Inattention to those pivotal issues suggests lack of strategic vision and forward thinking on the part of those in charge of formulating American foreign policy toward the Middle East.

Second, the civil war in Syria demonstrates both the United States' lack of vision and reactionary policy. Further, the interests of Iran and Russia converge into one cohesive strategy with a specific objective: supporting the al-Assad regime militarily, economically, and politically. Russia, for instance, embarked on a major security gamble in the Middle East as Putin made the strategic decision to move heavy weapons into Syria. While Russia aims its tactical short-term calculations at Syria, its long-term strategy goes beyond the borders of the civil-war-torn country. For instance, in support of its efforts to turn the geopolitical tables, Russia gained military momentum earlier on by holding joint naval military exercises with Iran. In August 2015, Russian warships, *Volgodonsk* and *Makhachkala*, docked in Anzali Port near the Caspian Sea.[61] The timing could not have been better as these naval maneuvers came only a few weeks after the nuclear agreement between Iran and the P5+1. This is significant because it demonstrated, as I argued in previous writings, that it would be just a matter of time before Iran and Russia entered into a strategic military alliance. This cooperation is getting stronger as both countries realize the need for each other despite having different objectives. For instance, Putin supports Iran in building new nuclear power plants using Russian technology.

The sale of Russia's S-300 advanced missile defense systems supports the argument that Russia is willing to go far to communicate to the West that the latter should start considering Russia's global interests. Moreover, Russia can no longer be overlooked when it comes to its geostrategic interests. Russia's ongoing engagement with Iran through joint military exercises and military activities in Syria supports my claim that Russia seeks to reassert itself in the Middle East. This is evidenced by the number of Russian troops and amount of military hardware presently in Syria. Putin's military support for the al-Assad regime continues to confuse the Washington establishment regarding his true intentions. Gabriela Baczynska writes, "The sources, speaking to Reuters on condition they not be identified, gave the most forthright account yet from the region of what the United States fears is a deepening Russian military role in Syria's civil war, though one of the Lebanese sources said the number of Russians involved so far was small."[62]

While Putin is adamant about the need for Russia to regain its lost global status and prestige, his strategy in the Middle East is not limited to the military apparatus. Rather, he uses a tactic wrapped in a diplomatic narrative that communicates to the West how cooperative Russia is in fighting terrorism. For instance, Putin claims that any political solution to the Syrian conflict must incorporate the al-Assad regime. Russia is not only providing military cover on the ground for the Syrian regime but also diplomatic cover in the international forums. This is no more evident than when, for instance, vehicles in Damascus are plastered with pictures of Bashar al-Assad on one side and Vladimir Putin on the other.

Given his awareness of how global dynamics work, Putin claims that his nation's military presence in Syria is to fight ISIS since the United States is unwilling to engage with ground troops and would rather count on its air superiority to destroy ISIS. This strategy, thus far, has been ineffective and has not achieved its intended objective: complete destruction of ISIS. Rather, it only limited the terrorist group's ability to acquire more territories.

Once again, Putin's strategic calculations and tactical maneuvers revived the tensions that once marked relations between East and West. I argued in previous writings that the Syrian quagmire and Iran's nuclear negotiations provided Putin the right platform on which to rearrange the chairs around the geopolitical table where attendees would sit. Could this be the reason Russia declined to disclose the scale and scope of its military presence and mission in Syria?

Putin is a step ahead, keeping the West guessing as to what his political moves mean. For instance, when he announced on March 15, 2016, the withdrawal of Russians forces from Syria, the West was quick to categorize it as an admission of failure. There are two things the West forgets to consider about Russia under the leadership of its strongman, Putin. First, never bet too heavily on what the former KGB strongman will do next; second, things are not always what they seem to be. Military analysts in NATO, Western think tanks, and pundits failed to conclude that Putin's announcement amounts to no more than a drawdown because, despite news bulletins to the contrary, Putin still maintains a military presence in Syria. Intelligence imagery indicates the Russian buildup of more advanced military capabilities that include but are not limited to a Sukhoi Su-24M bomber aircraft group, Sukhoi Su-30SM "Flanker" multirole combat aircraft, Su-25 "Frogfoot" strike aircraft, and many helicopters, including the Kamov Ka-52 "Alligator" attack helicopter.[63]

I argue that, indeed, Putin caught the West by surprise with his announced Syrian drawdown. But make no mistake: this partial withdrawal does not mean that things are settled militarily. The city of Aleppo remains partially encircled by jihadi forces who in turn are encircled by coalition forces headed by the Syrian regime. I ponder this question: What does Russia's drawdown mean for Syria, the Middle East, and geopolitics in general? We must consider four key points as we debate this unexpected turn of events.

First, the unexpected reduction in forces hardly suggests Russia's involvement in Syria is over. To the contrary, Russian military capabilities continue to grow. Russia's ongoing targeting of jihadi groups, mainly Jabhat al-Nusra (another name for al-Qaeda) and the Islamic State, suggests that the West misread Putin's intentions. Alastair Crooke, a former MI6 and author of *Resistance: The Essence of Islamic Revolution*, argues that the cessation of hostilities between the opposition and Syrian forces, was brokered by senior Russian officers. Yet, to quote a Russian military journalist: "The fleet remains; antiaircraft systems remain; the tanks remain; all the marines remain; the helicopters remain; some of the aircraft will remain. Only some of the aircraft and their service personnel are being taken out. And they can come back, of course, in the space of three or four hours."[64]

Second, could Russia's drawdown have an impact on the ongoing Geneva talks? Is Putin indirectly influencing the outcome of these talks? Or is he exposing the United States' lack of leadership in convincing its allies, Saudi Arabia, Qatar, and Turkey, to refrain from funding and arming their proxies or even marching defiantly toward Damascus? Answering these questions is no easy task. While the United States might have leverage over the desert kingdom and Qatar to bring about their change of course, it will find it a challenging proposition to do so with Turkey. The reason is that Turkish president Recep Tayyip Erdogan needs the Syrian conflict to drag on longer to keep his hold on power. One does not need to look far to see how Turkey is using the Syrian refugees' quandary to blackmail Europe to secure money and gain European Union membership.[65] Shame on Turkey, a Muslim country that calls itself a secular, democratic state.

Third, all along, Putin has wanted to put Russia in a strong position vis-à-vis not only the Syrian issue but also the conflict in Ukraine, tensions with the United States, and the ongoing sanctions by the United States and Europe on Russia over its annexation of Crimea. Whatever the outcome may be, Russia's latest move—and the rush by the United States to react to this drawdown—is not just about Syria but rather the future of the global order, or what passes for it these days.

The weapons deployed suggest that Putin's strategic interests go beyond the Syrian borders. Putin figured out that he could hit two birds (provide support to the al-Assad regime and send a message to NATO) with one stone (advanced heavy weapons). In doing so, he proved to the West that Russia still wields influence and can change the political geopolitical equation in the Middle East.[66]

Fourth, Iran's nuclear dossier has created anxiety within the region as countries including Saudi Arabia, Jordan, Turkey, and Egypt, among others, start to consider what to do if Iran gets the bomb. This anxiety about Iran's nuclear deal is no more evident than in the Arab world, especially the Gulf states. The fear lies not only in Iran's raising its military and economic power but also its religious reach, mainly its Shia ideology. The ongoing schism between Shiites and Sunnis is not new; rather it is an old, bloody contest that traces its roots to the early era of Islam, especially after the death of Prophet Muhammad. It is imperative to place the fear that Gulf monarchies and some Arab countries have vociferously expressed toward Iran's nuclear agreement within the context of what it means for their own survival. The presence of a Shia minority in Saudi Arabia (southeastern region), Bahrain, Kuwait, United Arab Emirates, and Qatar has always been regarded with suspicion. Some fear that one day they might rise up against the central government. When the Arab Spring swept the Muslim world, minority Shiites seized the opportunity to make their voices heard by joining in the demonstrations against tyranny and discrimination, but to no avail. Their attempt was met with a swift, strong response from the central governments of Saudi Arabia, Kuwait, and Bahrain, to name but a few.

While sectarian tensions between Sunnis and Shiites have always existed, Iran's nuclear agreement with the West pushed those tensions to the forefront. What has been left out of the debate, conveniently or ignorantly, is that the rising tensions in the wake of Iran's nuclear agreement are more about religious dominance than Iran's economic or military gains in the Middle East and beyond. For instance, Saudi Arabia continues to express vehement opposition to Iran's nuclear deal. Yet, the real issue, from my perspective, is Saudi Arabia's unwillingness to relinquish its religious dominance in the Muslim world. As result, I argue that all the statements coming out of the Gulf states, and mainly from Saudi Arabia, aim to categorize Iran's nuclear deal as a threat to peace and stability in the region. For instance, Tariq Al-Shammari, a Saudi analyst and president of the Council of Gulf International Relations, argues, "This agreement, from our point of view, represents an indirect threat to Gulf and Arab interests and peace."[67]

The ongoing tensions between Saudi Arabia and Iran are evidenced in none other than Syria and Yemen. Saudi Arabia, Turkey, and Qatar are funding and arming Sunni rebels in Syria against the al-Assad regime, a regime that Iran vows to support to ensure its survival at any cost. Similarly, in Yemen, Saudi Arabia has been leading a Sunni Arab coalition against the Houthis, Shia rebels supported by Iran. The objective is to restore to power the Saudi-backed Yemeni president, Abd-Rabbuh Mansour Hadi. To its detriment, the desert kingdom, thus far, has been unable to defeat the Houthis. Similarly, the ongoing sectarian violence in Iraq suggests that it is unlikely that the Sunnis will ever regain power in the land between the Tigris and the Euphrates.

All these upheavals suggest that the Middle East is now split along religious lines: Sunni versus Shiite. Therefore, the argument that Iran's nuclear deal paves the way for Tehran to dominate the Middle East and the Muslim world, for that matter, lacks sound judgment, credibility, and pragmatism. Frankly, it's nonsense. I argue that it shows more of a political convenience that aims at specific objectives (sale of weapons, access to energy sources, and so on). And here is the part that many in the West are oblivious to or misled about: the notion of Sunni unity is a myth. The Sunni world is more divided than ever.

Iran's nuclear agreement demonstrates a spike in empty rhetoric with a military tone to it. My reference is to the leading opposition force in the Middle East, Saudi Arabia. For instance, Prince Bandar, the long-serving Saudi ambassador to the United States, one who is familiar with the American political landscape, was quoted as saying, according to the *Times of London*, that the kingdom and other Gulf powers are willing to take military actions against Iran without the support of the United States.[68] While the Saudi statement could not be further from reality, it certainly makes a great headline and public relations statement through which the desert kingdom aims to project its strength.

Yet, conventional wisdom suggests that the kingdom could not conduct a successful operation as simple as a humanitarian or rescue mission. It has been dependent on the United States for its survival for the last seventy years and remains so. Similarly, the ongoing chaos in Yemen—a result of Saudi Arabia's poor and chaotic military execution—shows the desert kingdom's ineptitude when acting alone. The Saudi royal family should understand that engaging Iran militarily is not as easy as it seems. Iran is as strong militarily as the kingdom, perhaps even stronger. We must remember that Saudi Arabia is not the only country in the region that continues to express its concerns over Iran's nuclear agreement with the West. Israel has similar concerns, for different reasons.

I address Israel's position on the agreement separately. For now, suffice it to say that key regional countries, besides Israel, are noting Iran's buildup of nuclear capabilities, be it a power plant or other nuclear infrastructure. It cannot come at a worse time for the region as Iran and Saudi Arabia are engaged in proxy wars in Syria and Yemen. The conflict between Iran and Saudi Arabia, however, goes far beyond the nuclear issue; rather, it extends to religious dominance within the Muslim world; political, ideological, and economic influence in the Middle East; and the ability to challenge the status quo the Kingdom of Saudi Arabia has supported since World War II.

Fourth, the political outcome of Iran's nuclear deal with the West could follow a path similar to China's in 1972, when, under President Richard Nixon, the United States reached out to Beijing. Simply said, Iran could come out of the shadows to reintegrate within the global economy. Thus, it would challenge countries like Saudi Arabia in the energy sector given that Iran is the fourth-largest oil producer. Would the economic opportunities that await Iran to modernize its economy lead China, Russia, and the European Union to compete for access to the Iranian market? Economically speaking, we are already witnessing a tremendous shift. Lifting sanctions on Iran has already been welcomed with a huge business appetite. According to Iran's minister of transportation, Iran reached a tentative deal with the European consortium Airbus to buy 118 passenger planes. Further, Iran is looking to purchase four hundred passenger planes over the next decade.[69]

Political dynamics in other parts of the world made the situation even more complicated. Of interest is how Russia's annexation of Crimea on March 18, 2014, soured relationships between the United States and Russia. One wonders how, subsequently, Russia and the United States could sit on the same side of the table, across from the Iranian negotiators, moving the negotiations forward in the hopes of reaching an agreement on Iran's nuclear program. Worse yet, the United States decided to impose sanctions on Russia to punish it for annexing Crimea. In a diplomatic retaliation, Russia started to veto any proposition introduced by the United States or any other Western members of the UNSC advocating for additional sanctions on Iran, thus undermining US efforts during the negotiations.

The Iranian negotiating team considered many geopolitical factors. The team's thinking provided Iran a justification to drag out the negotiations given the United States' entanglement in the never-ending spiral of conflicts in the Middle East—including the civil war in Syria, the rise in the sectarian violence in Iraq, and, later, the emergence of ISIS. Those force-producing events encouraged new thinking within Iranian political elites, who persuaded the negotiating team that, in turn, Iran could assist the United States in its efforts to defeat ISIS. However, the possibility of such cooperation comes at a price: in return, Iran expects flexibility regarding its controversial nuclear program.

While these developments shape the security architecture of the Middle East, Iran and Russia continue their partnership in the hope of developing military cooperation. Tense US-Russia relations over the latter's annexation of Crimea pave the way for Iran and Russia to move rapidly into a more inclusive partnership. This stems from the following: (a) Iran would like to possess advanced missile systems, besides the domestically produced ones, to project strength in case Israel decides to attack its nuclear facilities; and (b) Russia would like to prove to the West that it can still tilt the political balance by changing the geopolitical equation in the Middle East.

The Path Forward

Iran's nuclear dossier has been on a roller coaster given not only the division it created among major and regional powers but also the opportunities it has presented. Iran's nuclear dossier exposed the hidden agendas of major powers when it comes to the security of strategic interests. Iran's nuclear program demonstrates that, despite having a common objective: preventing Iran from acquiring nuclear technology, major powers are disjointedly pursuing their strategic interests. For instance, Russia and China want to grow their presence in the region while engaging regional countries in military and economic ventures. Simply said, whatever strategic interests the United States hopes to achieve indeed differ from those of China and Russia, for instance. I illuminate this salient theme in the hope of challenging the theories and assumptions of other scholars and pundits.

Differences among major powers about Iran's nuclear dossier amount to strategic interests. Russia is of interest in this debate, and why is that? The answer lies not only in the speed of unfolding events but also whether Russia seizes the opportunity and used Iran's nuclear program to pursue an assertive foreign policy in the region. This insight is gained not from delegating from an armchair but rather from being on the ground, evaluating events as they unfold.

Within this context, I argue that, in the case of Russia, Moscow had to quickly adjust its position on Iran's nuclear dossier. Yet, the thinking among global-affairs analysts suggests that Iran's nuclear dossier changed the geopolitical calculations among major powers for many reasons. Chief among them is the hostile relations between Iran and the United States. Further, Russia and China realized the political opportunity to steer their separate strategic interests through Iran's negotiations with the West over its nuclear program. Given those developments, questions emerged about whether Russia, for instance, would be in a position to influence the process and outcome of the nuclear negotiations. The consensus among academics and international security analysts is that Russia realized that Iran's nuclear dossier would eventually become the main security issue and that no one particular country among major powers would have the sole power and authority to decide the outcome of Iran's nuclear program. In a way, Russia played its political cards intelligently, quietly influencing the nuclear negotiations to its favor. On some occasions, Russia displayed its alignment with the West. Other times, it exerted its influence when it vetoed additional sanctions on Iran. Russia's strategy removed any ambiguity over its long-term objective.

Consider this argument from a different perspective: the assertion that Russia cannot act independently from Washington when it comes to Iran's nuclear dossier is not merely inconclusive—it is flat-out wrong. Among those who suggested this thinking is Alireza Noori, an expert on Russian affairs. He writes,

> Although, in accordance with the logic of international political equations, Russia cannot behave totally independent when it comes to Iran's nuclear case, and it is logical for Moscow to get aligned with Washington in order to gain more benefits, its "excessive" dependence on the "United States" steering role in this case is, without a doubt, an important shortcoming for Russia. Even the general course of the nuclear case has proven that Russia is by no means able to take the initiative outside a framework which has been set for it by the United States. "Practical" inattention to Moscow's step by step initiative for

finding a solution to Iran's nuclear case was a clear indication of how limited is Moscow's maneuvering room on this issue.[70]

I find this assertion troubling because it was Russia that influenced the nuclear negotiations over Iran's program to its favor. Behind closed doors, Russia convinced Iran to reach a deal with the West. Russia understood that when the West and Iran reached an agreement, Iran's economic sanctions would be lifted; hence, Iran would generously reward Russia for its support. As of this writing, Russia and Iran are teaming up in the Syrian theater to support the embattled Syrian president Bashar al-Assad to ensure he remains in power.

US-Iranian Relations before the Revolution of 1979

In the broad spectrum of history, events come and go; some leave their mark, while others fade away. The events that have a tremendous global impact tend to influence policies and shape international relations. Iran fits this description. In a matter of one year, 1979, the Islamic revolution in Iran changed the entire history of US-Iranian relations for decades to come. The relationship went from cooperation and collaboration to animosity, mistrust, and hatred.

A plethora of information has been written about this epoch, one marked by the complete influence, or for lack of a better term, the excessive power of the United States over the Iranian government. But before delving into how this influence made possible resentment and anti-Western, especially anti-American sentiment, it behooves us to examine US-Iranian relations from the early 1950s to 1979, the year of the Islamic revolution that ended US influence over Iran once and for all. This resentment not only impacted the diplomatic relations between the two countries but also halted Iran's peaceable relations with the United States.[71]

The United States' dominance over Iran's political landscape emerged when the United States prepared Mohammad Reza Shah to assume power over his rival, Mohammad Mossadegh, a charismatic Iranian premier. The United States' support came through a coup orchestrated by the CIA and British MI6, code-named "Ajax."[72] While the reason for the coup was mainly Mossadegh's decision, in 1951, to nationalize the Anglo-Iranian Oil Company, Britain and the United States feared at that time that such a move would reduce their regional influence. The other reason pertained to the United States' fear that Iran might fall into the communist orbit, thus inviting the presence of the USSR into the Persian Gulf. Those motives prompted US president Dwight Eisenhower to authorize the coup. Once Mossadegh was removed from office, the United States made sure that the reinstatement of the shah remained intact. To accomplish this objective, the United States threw its support behind General Fazlollah Zahedi, who pushed a US and UK agenda. Thus, the United States' control of the Iranian government began.

US-Iranian relations in that era flourished under a false premise that, as we now witness, failed to withstand the passage of even a few years. The Iranian people disapproved of this new relationship. The shah promoted the idea that Iran and the United States enjoyed a great relationship. But, in reality, that was far from the truth. There was discontent among Iranians. Animosity grew toward the shah, and hatred and resentment fomented toward the US. To maintain a firm grip on power, and with the support of the United States, the shah embarked on repressive policies managed and guided by the Iranian National Intelligence and Security Organization (known as SAVAK). Interestingly, the United States created and supported this Gestapo-like organization. The organization took it upon itself to be the judge, prosecutor, and jury. It embarked on a campaign of arrests, torture, and killings of thousands of the shah's opponents under the watchful eyes of the United States.[73]

During this era in Iran, the gap between the haves and the have-nots started to widen. The reason is that, with a few other families, the shah controlled most of the wealth with links to the oil industry. The foci of the United States and Britain remained on having access to oil sources in Iran and ensuring that Iran served as a bulwark against the USSR. As a result, the United States turned a blind eye to the atrocities the shah committed toward Iranians that he perceived as his enemies since they not only viewed him as a puppet of the United States but also disagreed with his policies.

During this era, the Arab world was going through a political shake-up ranging from military coups and economic stagnation to the rise of Islamist parties and terrorism. These developments, mainly their religious aspects, introduced paranoia among the US foreign-policy establishment toward Islamist parties and Islam itself. Why would that interest the United States within the context of US-Iranian relations? The answer lies in the United States wanting to have some form of secular reforms in Iran (a) to avoid having the United States be perceived as the one in the driver's seat of Iran's internal affairs (dictating policies), and (b) so the shah would be presented as progressive in the hopes of his popularity surging.

Those efforts failed miserably. How could such plots succeed when the shah, with the approval of the United States, committed crimes against innocent civilians who happened to disagree with his policies? US-Iranian relations were unbalanced, tilted to favor US interests. Could that explain why the United States ignored the atrocities the shah committed while helping him build his military apparatus? This is evidenced by how the high echelon of the US government, including US presidents, ensured that the passage of legislation maintained good relations with the shah rather than Iran. The shah did the same by passing regulations through parliament that not only pleased the United States but were also applied according to Washington's dictates.

As stated earlier, the United States' relations with Iran before 1979 focused mainly on ensuring US hegemony in the region. Yet, the perplexing aspect of the United States' foreign policy toward Iran at that time was that it seemed to be on two opposite trajectories, promoting democracy and human rights on one hand while turning a blind eye to atrocities and abuses on the other. This is exactly what happened when the United States endorsed the Iranian government, despite the shah's human rights abuses, for its democratic process and human rights record. Such hypocrisy on the part of the United States!

It is as though we have not learned any of history's lessons, and so, history repeats itself. The military coup in Egypt, in 2013, that saw the removal of the democratically elected president Mohamed Morsi comes to mind. Yet, the United States decided to turn a blind eye to this like it did in Iran four decades earlier. By doing so, the United States sent the wrong message to the world about the rule of law and the American democratic system.[74] It is almost as though US double-standard behavior is part of its foreign-policy genetic makeup. Why is the US foreign-policy establishment unwilling to apply the lessons of history when formulating its strategy? It is beyond me. It defies logic. I find myself asking, for instance, if US policy makers considered the possibility that youth in countries like Iraq and Afghanistan (the former we invaded in 2003; the latter where we still are after fifteen years of war) might rise up when they got older to seek revenge for the destruction the United States caused in their countries? It is a fair question. I recall during my tours in Afghanistan, Yemen, and Iraq how young people stared at me, rage emanating from their eyes. The Iranian case is no different because the hatred the Iranians harbor today toward the United States results from the latter's support for the shah.

The late 1950s and early 1960s saw the emergence of opposition in Iran in response to US support for the shah's repressive regime. Ayatollah Khomeini opposed the shah and set US-Iranian relations on a collision course. The United States made two major mistakes in the early 1960s regarding Iran. First, the Kennedy administration pressured the shah to introduce a bill that allowed non-Muslims to be political candidates. Second, under pressure from the United States, the shah gave the Americans stationed in Iran immunity from prosecution in Iranian courts. It was the latter legislation that gave Khomeini the justification to voice his opposition to how the United States was not only influencing Iran but also corrupting it. Thierry Brun writes, "If any of them commits a crime in Iran, they are immune. If an American servant or cook terrorizes your source of religious authority in the middle of the bazaar, the Iranian police does not have the right to stop him. The Iranian courts cannot put him on trial or interrogate him. He should go to America where the masters would decide what to do. . . . We do not consider this government a government. These are traitors. They are traitors to the country."[75]

Khomeini's jailing and subsequent expulsion from Iran marked the beginning of his open opposition to the shah, which eventually manifested itself in the 1979 revolution. Unable to change the political, economic, and religious realities on the Iranian landscape, the United States found itself watching from the sidelines as different segments (women, students, and religious reformers) joined Khomeini's cause. Equally important, university students inside and outside of Iran participated in these events. The activities aimed at exposing the shah's failed social and economic policies, policies that had impacted Iranians of all stripes, except for wealthy families who started to realize that the beginning of the end for the shah was near.

The massive crackdown undertaken to suppress demonstrations ultimately paralyzed the country. Once again, the United States could do nothing about such force-producing events. At that juncture, Khomeini increased his verbal attacks on the shah, calling for his departure. Khomeini thereby gained confidence in his ability, fueled by popular support, to rid Iran of the tyrant, the shah. Regardless of the presidential administration that it served at the time, the United States' foreign-policy establishment failed to realize that supporting the shah in committing atrocities and violence against human dignity would eventually catch up with the shah and register on the moral compass of those in the United States who formulated and authorized that policy. It exposed the United States' double standard and the bravado about democracy, which amounts to nothing but empty rhetoric. It also exposed the United States as a shallow nation at the core, a truth many in the United States did not want to hear.

In other words, the US-Iranian relationship was based on an unsound foundation. I understand the argument that in global affairs a country's own interests come first, and I support that. However, if the United States is to build long-lasting relationships with other countries, it ought to look inward first to identify and confirm its own principles before advocating them to others. The current US-Iranian relations expose the inherited flaws within our foreign-policy establishment. Policy changes ought to precede chaos. As Kenneth Pollack argues, "As anyone who has served in the government knows, it is easiest to redirect the ship of state if changes are made well in advance of dramatic events—and hardest when forced to react to sudden, unforeseen developments."[76]

Today, the United States is faced with some major decisions regarding Iran: Are we going to make a path toward normalizing relations or stay the course of hostility and mistrust? Will the United States engage Iran militarily if the latter develop nuclear weapon capabilities? Or will the United States cooperate with Iran when their goals (defeating Sunni militants) overlap? I intend to answer these questions in the subsequent chapters. Suffice it for now to state that after 1979, US-Iranian relations took a 180-degree turn for the worse. Yet, I maintain a hopeful, though pragmatic, outlook that change can be achieved if both countries are willing to sacrifice. Time will tell, sooner rather than later, I hope.

US-Iranian Relations after the Revolution of 1979

Much can be said about the United States' and Iran's relations after the Islamic revolution of 1979. Scholars, foreign-policy experts, and global-affairs analysts agree that animosity and tension in the American-Iranian relations emerged forcefully after Iran's Islamic revolution of 1979 and subsequent hostage crisis.

Undoubtedly, following the ouster of the shah, a close US ally, and the establishment of the theocratic ruling of Khomeini in the aftermath of his 1979 revolution, relations between the United States and Iran nosedived. I cannot recall a
similar situation in which the United States felt such animosity and tension toward another foreign government, at least not since the end of World War II. The
level of mistrust from both Tehran and Washington, evidenced by their vitriolic
day-to-day exchanges, have deeply affected both countries. From the US perspective, Iran's attempts to undermine US geostrategic interests in the region hinder
the United States and prove politically inconvenient. From Iran's perspective, the
reasons for the breakdown of the relationship are both economic and political.
The sanctions the United States imposed on Iran following the revolution of 1979
choked Iran's economy. Yet, Iran managed to survive by using its own resources,
in addition to the black market, to circumvent the sanctions. Politically, Iran
takes advantage of the United States' policy miscalculations in the region. US
engagement in Lebanon in the 1980s and its invasion of Iraq in 2003 are the products of such miscalculations.

The United States' policy toward Iran after the 1979 revolution continues to
drive US policy makers of all political stripes to maintain, and further impose,
tough sanctions on Iran. The reason is that, despite the sanctions, through its
proxies Iran can influence political outcomes on the ground. This is no more
evident than during the US invasion of Iraq in 2003. One does not have to be a
political genius to see how Iran succeeded in influencing events in Iraq to Iran's
political satisfaction. Could a similar argument apply to the civil war in Syria
and upheavals in Yemen? This indicates how geopolitical shifts can have a tremendous impact on global affairs and international relations. In 1979, virtually
overnight, Iran went from being a strong US ally and a major pillar of US foreign policy in the Middle East to a pariah. During his administration, former US
president George W. Bush referred to Iran as a member of the "Axis of Evil."[77]

Whatever the case may be, Iran's geopolitical aspirations in the Middle East
remain of great concern to the United States and its allies in the region. Now
that Iran has reached an agreement with the West over its nuclear program, the
current question is what kind of Iran do the United States and the West expect?
Will Iran change its behavior and restructure its foreign policy? Or will we see an
aggressive, confident Iran interpret its agreement with the West as a victory? Will
the support of Russia and China help Iran challenge the status quo not only in
the Persian Gulf but also in the greater Middle East? How will main allies of the
United States in the region (Saudi Arabia, Jordan, United Arab Emirates, Qatar)
react to Iran's growing influence? These questions are at the heart of US foreign
policy moving forward. The grandiose plans that the Washington establishment
made during the shah era faded the moment he was overthrown. Based on a 1974
report issued by the US embassy in Tehran, Donette Murray, a senior lecturer

of defense and international affairs at Royal Military Academy Sandhurst in the United Kingdom, argues,

> [The] US for its part has great stake in Iran's survival and welfare because (A) it has ability and willingness to play a responsible role in region; (B) it has history of cose [sic] and friendly ties with the US; (C) it is reliable and important source of oil and other resources, (D) it is growing market for our goods and services ($7 billion in US civilian and military contracts in past Two years) and a hospitable location for US investment; (E) it provides essential air corridor between Europe and Orient, and (F) it allows us to use its territory for special communications and intelligence facilities . . . in recent years and as our aid and tutelage phased out a close relationship as equal partners has evolved.[78]

Alas, these opportunities are gone. In reality, US-Iranian relations took a different trajectory, one marked by hostilities, mistrust, and animosity. While each US administration approached US-Iranian relations differently (each one wants to leave its own mark when it comes to Iran), they also share a commonality: they see Iran as a pariah. Despite all this, there were times when doors of communication between the two nations were secretly kept open. Nevertheless, it is vital to provide a snapshot of how each American president after Iran's revolution of 1979 approached US foreign policy toward Iran.

The Carter Administration

The presidency of Jimmy Carter, thirty-ninth president of the United States, was influenced not only by what he inherited from previous administrations but also by what transpired after he took office. The restoration of the shah to power, through US and British covert operations, made it possible for the United States to play a greater role in the region while Iran protected US interests in the Persian Gulf. However, the Carter administration found itself in a political downward spiral as the Islamic revolution in Iran was unfolding in 1978. Carter's foreign-policy team became aware of the challenges awaiting the administration as it appeared inevitable that the shah would eventually be overthrown.

Once the hostage crisis made the 1979 revolution a reality, Carter's administration found itself paralyzed, unable to adapt quickly to these fast-changing events. These force-producing events suggest that Carter's foreign-policy team failed to grasp what those challenges entailed. Under no circumstances am I suggesting that that failure should be interpreted as an intelligence letdown; rather, branches of the US government failed to interpret and share intelligence information. Gary Sick, principal White House aide for Iran and the Persian Gulf on the Carter administration's National Security Council, provided a similar argument: "The Iranian revolution and the hostage crisis dramatized to U.S. policymakers the gap between U.S. regional interests and its ability to project force."[79]

It is fair to say that while Carter tended to handle the Iran crisis, mainly the hostage situation, with sensitivity given the high stakes (US hostages might be killed), his failed attempt to rescue the hostages cast doubt on his ability to confront and use military force when faced with conflict. Unfortunately, the failed rescue attempt was the basis on which he lost his reelection bid for the White House to Ronald Reagan.

The Reagan Administration

A host of issues marked Ronald Reagan's presidency. Some were good; others were bad. However, historians agree that the release of the US hostages held in Iran following Reagan's election set the United States on a different trajectory from that of his predecessor, Jimmy Carter. The release of the hostages, the culmination of months of secret negotiations with Iran, allowed Reagan to push forward with his campaign promises (reduction of taxes, increased defense spending, and an end to the Cold War). Reagan's approach to foreign policy was guided by the mantra "peace through strength." The aftermath of the 1979 Iran revolution played a major role in shaping his policy toward the Middle East. For instance, Reagan ordered naval escort in the Persian Gulf to ensure the free flow of oil while the Iran-Iraq War raged. Simultaneously, he decided to support anti-communist groups, including those in Asia, Africa, and Central America.

With this in mind, I discuss two issues in this brief account of Reagan's presidency as it relates to Iran: (1) the role his administration played during the Iran-Iraq War and (2) Reagan's decision to sell weapons to Iran in return for funneling money to Nicaragua. Regarding the Iran-Iraq War, Reagan's administration sought to assist Iraq in the hope that it would defeat Iran. However, when it became apparent that Iraq would be unable to defeat the United States' sworn enemy, Iran, the Reagan administration worked behind the scenes to end the conflict, since dragging it out would undermine US interests in the region. However, it would do the readers an injustice if I were to omit key information that led to the Iran-Contra affair. The essence of this information convinced the Reagan administration to sell weapons to Iran through Israel in return for the release of the US hostages in Lebanon held by Hezbollah, a proxy of Iran.

The Iran-Contra affair was a clandestine operation conducted without authorization from Congress. The millions of dollars acquired from the sale of these weapons were funneled to a right-wing "Contra" group, opponents of the Nicaraguan Sandinista government. At the heart of the matter is not that the Reagan administration intended to prevent communism from taking hold in Central America. Rather, it was the administration's desire to prevent the Iran-Contra affair from exposing the illegality and abuse of executive power. Malcolm Byrne writes, "In fact, the Iran-Contra affair was about much more than that single question. It encompassed highly dubious, and possibly illegal, acts with respect both to strands of policy and to broad-gauged attempts to cover up administration activities following their public disclosure. Far from being the work of a few mid-level 'rogue operatives,' it involved at various stages an array of senior officials including the president and vice president themselves. But by presenting the diversion to the public in such a dramatic way, the officials managed to marginalize the importance of other critical elements."[80]

The Reagan administration sought to make its political mark and change the political and historical trajectory by setting the agenda for reshaping the region's political landscape. This explains why, during the Iran-Iraq War, Iran received support from countries including Syria, Libya, and China as well as covert arms transactions from the United States during the presidency of Ronald Reagan.

The George H. W. Bush Administration

If there is one thing to be said about George H. W. Bush's presidency, it is that he was elected during a tumultuous time marked by rapid historical changes on the global stage. Chief among them was the end of the Cold War—a war that defined international relations between West and East and manifested on the global stage for four decades. The collapse of the Soviet Union helped the United States become the dominant power as Bush shaped the global political landscape as president. Bush indicated such work as he famously referred to as shaping the "new world order."

Yet, the emergence of the United States as the only superpower did not alter the decades-old animosity between Iran and the United States. Ongoing tensions and mistrust were even more prevalent as Hezbollah, a proxy of Iran, was involved in the kidnapping of American and other Western hostages in the 1980s. The role Iran played in this issue and others, including the support it provided to the Palestinian Islamic Jihad and Hamas, convinced the Bush administration that the Middle East would continue to draw in the United States more than ever.

The Bush administration assumed that, after the end to the brutal eight years of war between Iran and Iraq, political tensions in the region would subside. But how could they when Iran's use of terrorist organizations (the Khobar Towers bombing, the kidnapping in Lebanon) to further its agenda raised serious concerns in the Bush administration? So, the Bush foreign-policy team's assumption that leaders of Iran and Iraq would eventually begin the slow process of rebuilding their countries was misguided.

To its credit, the Bush administration worked behind the scenes to free the hostages held in Lebanon. It worked through the United Nations rather than directly contacting the Iranians, especially after the embarrassment of the Iran-Contra affair. After the release of the last American hostage in 1991, two years after the death of Ayatollah Ali Khamenei, the possibility of opening a dialogue with Iran in the hope of improving relations had passed. One particular factor that contributed to this outcome, in my opinion, had to do with Iran's hopes that the release of the hostages would convince the United States to relax some of its sanctions placed on Tehran. The Bush administration found itself getting nowhere with Iran and decided that it might as well apply tougher sanctions pursuant to those of previous administrations. Keep in mind that Iran experienced chaotic internal political dynamics at that time, especially after the June 1989 death of its spiritual leader, Ayatollah Khomeini.

While the focus of the Bush administration remained on how to deal with the growing influence that Iran wielded through terrorist activities and sabotage, other events in the region took the administration by surprise. Consider Iraq's invasion of Kuwait in August 1990. Bush wasted no time in aggressively claiming America's greatness and leadership. He assembled an international coalition to drive Iraqi forces out of Kuwait because doing nothing presented a far greater risk.

From the Bush administration's perspective, the concern went beyond just allowing an aggression to go unchallenged. The implications for global economics were significant given that Kuwait sits on major oil reserves. If Saddam Hussein's aggression were allowed to stand, an American diplomat noted at the time, "he would control the second- and third-largest proven oil reserves with the fourth-largest army in the world."[81] It could set a precedent for more aggressions in the region. There was also concern about how Iran might perceive Iraq's aggression if the United States failed to intervene. As a result, the Bush administration wanted to send a message to Iran and the region that the United States considered its commitment to the security of its allies in the region and stability in the Middle East a priority.

Regrettably, the United States' intervention also served Iran in a way that the latter could not have achieved on its own. The economic sanctions the United States imposed on Iraq crippled its economy and rendered its military weak. The outcome provided Iran an opportunity to engage in even more terrorist activities and meddle in the internal affairs of other Gulf states. When the United States invaded Iraq in 2003, critics argued that the United States should have decimated Iraq during the first Gulf War after it drove Iraq out of Kuwait. Yet, the Bush administration at that time (during Gulf I) decided not to march to Baghdad for a reason: weakening Iraq would be a strategic mistake. If Iraq maintained enough strength, the Bush administration argued, it could act as a bulwark against Iran's aspirations in the region. Simultaneously, the Bush administration feared that a strong Iraq could challenge, even invade, other neighbors as it did in Kuwait. The challenge for Bush was how to have a strong—but not too strong—Iraq to keep Iran from expanding.

Historians agree that the Bush administration's strategic vision was on the mark when it decided not to open lines of communication with Iran. Events on the ground, such as the Shiite uprising against Saddam Hussein, justified the administration not providing support to the Shiites, since Iran encouraged the uprising while providing support and logistics. Despite Hussein's ability to quell the demonstrations, Washington was concerned about the long-term impact of the influence Iran was gaining. Iran's behind-the-scenes involvement in the on-going upheavals—including the civil war in Syria, turmoil in Yemen, sectarian violence in Iraq, ISIS, and tension between Saudi Arabia and Iran—supports the thinking of the Bush administration.

The Clinton Administration

The collapse of the Soviet Union in December 1991 freed the Clinton administration from worries since it faced fewer constraints from the fallen superpower. The disintegration of the USSR made it possible for Clinton's foreign-policy team to focus on more pressing issues, including Iran's slowly but steadily increasing influence in the Middle East.

Unlike its three predecessors, Clinton's foreign-policy team was under no illusions regarding how it should deal with Iran. While there had been calls for the Clinton administration to reconsider its engagement with Iran, the administration ignored those critics. During this time, the so-called "moderates" in Iran were hoping that the Clinton administration would take a different path from its three predecessors, namely one of reconciliation, thus unfreezing some of Iran's assets in the United States. It is worth noting that, during this time, Ali Akbar Hashemi Rafsanjani, the elected president (1989–1997), tried to revive US-Iranian relations but to no avail. He hoped that the United States would show a gesture of goodwill by lifting some of the sanctions. Rafsanjani went even further by offering lucrative contracts to US oil companies. All these efforts were unsuccessful as the Clinton administration marched forward to make sure that Iran stayed isolated. Peter Rodman writes, "Christopher singled out Iran as the 'most worrisome' of a number of 'dangerous states' fueling regional tensions. He appealed to the Europeans to join us in a policy of firm containment and economic pressure on Iran, including denial of militarily useful technologies."[82]

Moreover, the Clinton administration rejected rapprochement with Iran because of mounting domestic pressure signaling opposition to any deal with Rafsanjani's government. US political interest groups' vehement opposition to rapprochement with Iran ended Rafsanjani's aspirations of being the Iranian president to reestablish ties with the United States after the Islamic revolution of 1979. After the election of Rafsanjani's successor—President Mohammad Khatami, a reformist—President Clinton had to shift his attention and priorities to domestic issues, thus preventing him from responding to Khatami's strongest signals yet for a change in US-Iranian relations.

The George W. Bush Administration

While the presidency of George W. Bush was marked by the events of 9/11 and the war on terrorism, the era also witnessed a deterioration in US-Iranian relations. Undoubtedly, the attacks of 9/11 contributed to the Bush administration labeling Iran as one of the powers of the "Axis of Evil." Subsequently, hawks in the Bush administration were compelled to craft a policy toward Iran harsh in tone and heavy handed in deeds (tougher sanctions). The invasion of Iraq in 2003 also contributed to that hawkish approach given that Iran was undermining US efforts to stabilize Iraq after the invasion. The policy reached a dangerous level when it became evident that Iran supported militias not only targeting US forces in Iraq but also pitting Shiites against Sunnis. Iran sought to widen the conflict between Sunnis and Shiites so the Muslim world would blame the United States for inciting tensions because of a senseless invasion.

Iran's nuclear program also exacerbated tensions between Iran and the United States during the George W. Bush administration. Despite the Bush administration's application of unilateral and multilateral sanctions on Iran to halt the enrichment of uranium, those efforts failed to completely halt Iran's nuclear program. I argue that Iran understood that the United States was stuck in Iraq given that the invasion turned out to be a disastrous venture for the United States' effort to spread democracy. Thus, Iran did not take seriously the United States' threatening language of regime change. The other factor was that Iran was convinced that the Bush administration was unlikely to engage in another military conflict in the Middle East given the American public's opinion about the wars in both Iraq and Afghanistan, which saw massive losses in blood and treasure.

The George W. Bush administration ended its second term with tensions between the United States and Iran at their worst level. It was not until the election of President Obama that the need to develop a meaningful dialogue and a realistic framework to address Iran's concerns became a sine qua non for at least opening dialogue, no matter how basic, between Washington and Tehran.

The Obama Administration

As of this writing, President Obama still has eight months before his second term is over. Historians disagree about what his legacy will be, years from now. Some of his campaign promises have been fulfilled, while others face major challenges due to the complex nature of global affairs. Critics argue that President Obama's decision to withdraw US forces from Iraq triggered the emergence of ISIS. The lack of a decisive strategy toward the civil war in Syria allowed the conflict to go global. The severing of diplomatic relations between Saudi Arabia and Iran is blamed on President Obama's agreement with Iran over the latter's nuclear program. Iran's nuclear program, I believe, will define Obama's presidency, at least from a foreign-policy perspective.

Critics argue that President Obama is naïve to think that an American foreign policy that promotes engaging enemies and pariahs will serve US interests. They argue that President Obama's naiveté is dangerous and could further the United States' leadership decline on the global stage. While this criticism was echoed in response to Obama's speech in Cairo, critics argue that the speech was heavy on rhetoric and light on substance. While Obama addressed his speech to the Muslim world, he aimed one of its main themes at Iran. Obama hoped his proposed strategy of engaging Iran could convince his critics that a dialogue with Iran was worth pursuing. As reported in the *New York Times* on June 4, 2009, Obama states,

> This issue has been a source of tension between the United States and the Islamic Republic of Iran. For many years, Iran has defined itself in part by its opposition to my country, and there is indeed a tumultuous history between us. In the middle of the Cold War, the United States played a role in the overthrow of a democratically-elected Iranian government. Since the Islamic Revolution, Iran has played a role in acts of hostage-taking and violence against US troops and civilians. This history is well known. Rather than remain trapped in the past, I have made it clear to Iran's leaders and people that my country is prepared to move forward. The question, now, is not what Iran is against, but rather what future it wants to build.[83]

Obama's approach to foreign policy traces its roots to a particular time frame: a 2007 Democratic primary debate in Charleston, South Carolina, where then-candidate Obama confirmed his willingness to negotiate with hostile leaders if they were also willing.[84] Fast-forward: after President Obama won the election of 2008, he made it a priority to find a way to jump-start talks with Iran. Of course the negotiations were not about reestablishing diplomatic ties. Rather, they were confined to addressing Iran's nuclear program. Without my detailing that aspect—since I address that theme in great depth in subsequent pages—suffice it to say that Obama took a gamble that, thus far, has proved somewhat successful in halting Iran's nuclear program. It will be interesting, in the coming years, to see whether Obama made a strategic mistake when he reached a deal with Iran over its nuclear program or paved the way for the cessation of hostilities from both ends: Iran and the United States. Time will tell—sooner rather than later, I hope.

To conclude this section, I argue that, while US-Iranian relations will most likely stay their current course, both countries might one day share common interests on which cooperation between the two hostile nations can exist. I say this within the context of how, since the demise of the USSR, the American foreign-policy establishment has tried to reinvigorate itself given force-producing events on the global stage. Over the last two or three decades, the national conversation has centered on the direction of US foreign policy. Questions emerged, for instance, as to where the United States was headed. Was there a foreign-policy strategy in place? How was the United States going to lead in a chaotic, soon-to-be-multipolar world? And how was the United States going to deal with countries that not only do not share its vision but are also armed with nuclear weapons? My reference in that final question is to Iran. US-Iranian relations have been marked by US culpability, particularly regarding the CIA's role in the overthrow of Mossadegh. This narrative continues even today to shape policies in both countries and influence US-Iranian relations in a way that allows each country to write the narrative that serves its interests and appeals to its populace.

If I have learned one thing in international relations, it is that context matters. I am concerned about whether President Obama's outreach to Iran could be construed in a different light. Said differently, Iran may interpret US outreach during the nuclear negotiations as America's way of confessing to its sins against the Islamic republic. The danger of this thinking is that it allows the Islamic republic not only to hold on to its revolutionary principles (key ingredients needed to maintain the power of theocracy) but also to claim the moral high ground. The forgone conclusion among theocrats in Iran is that the nuclear agreement serves as a victory over the United States and the West. That thinking will embolden Iran to move forward down the road to acquire a nuclear bomb.

The United States should develop a new way of thinking about its foreign policy toward not only Iran and the Middle East but also the globe. For this reason, US the foreign-policy cadre must include diversity in its ranks and a mixture of thoughts, vision, cultural background, ethnicity, and ideas. The world is changing, and American foreign policy should change with it. Ray Takeyh writes,

> Over the past 25 years, a major preoccupation of foreign-policy elites has been to forge a new grand strategy for the US. Scholars and practitioners tend to see a foreign policy adrift after the fall of the USSR, when containment of the Soviet Union's expansion became obsolete overnight. Seeing no major ideological or military rival, some believed the Owl of Minerva had taken flight, and that the end of history had reduced the need for strategic thinking. Alas, that fantasy came crashing down along with two big towers almost 15 years ago. Again, foreign-policy elites searched for a new strategy, this time for the age of Islamic terror.[85]

I am convinced that a nuclear Iran will eventually put American foreign-policy strategy—assuming there is one—to the test. Only then can we evaluate whether progress has been made in moving US foreign policy from a Cold War mentality to an era of mullahs and rogue states armed with nuclear weapons.

The International Community's Position on Iran's Nuclear Program

To some extent, the West—and the world, for that matter—was relieved to learn that, on behalf of its negotiating partners (P5+1), the United States had reached an agreement with Iran over the latter's nuclear program. Marked by ups and downs, the negotiations took about two years. There were times when both Iran and the P5+1 members walked away from the table. At other times, they were close but unable to reach an agreement. Once an agreement was reached in the summer of 2015, views of the deal varied greatly. The reason is that Iran's nuclear program is not just about Iran wanting to acquire the bomb and change the political calculus. Rather, it is about the nature of international relations, the contentious geopolitical issues, and the new political landscape that goes beyond the borders of the Middle East.

Before detailing what those views consist of, it is imperative to provide a road-map, so to speak, of this section. The map encompasses not only Iran's or the United States' views on the issue but also the views of other key players, like the European Union, China, Russia, and Israel, among other stakeholders. Readers must understand that Iran's nuclear program is a geopolitical story, an economic story, a political-identity story, an international-relations story, and an ideological story that transcends and defies the expected political rationale when dealing with global affairs. This section provides perspectives on Iran's nuclear program and the agreement reached with the West from various stakeholders' points of view.

The Iranian Viewpoint

To understand Iran's perspective on its nuclear program, one must first learn about the historical framework that made it possible for Iran to reach its historic agreement with the West over its nuclear program. The starting point of Iran's nuclear path goes back to the late 1960s, when the shah argued that, given Iran's vast petroleum resources, it was vital for Iran to use that commodity not only to produce electricity but also to engage in advanced energy and technological products. Therefore, I argue that Iran's nuclear aspirations are not new. Rather, they go back decades, to when the US intelligence community learned that Shah Mohammed Reza Pahlavi had embarked on clandestine activities to develop a nuclear program.

According to the Washington-based Center for Arms Control and Non-proliferation, the program consisted of extracting plutonium, a key component for nuclear weapons, from spent fuel using chemical agents.[86] That suggests that Iran's love affair with the nuclear program recalls the era of the shah, who saw an opportunity for Iran to use its massive petroleum reserves to realize far bigger goals. Was the shah driven by an egotistical desire to join the nuclear club? Or was he driven by a desire for global prestige? The answer suggests both. Roland Flamini, a former Washington-based chief international correspondent at United Press International and now a foreign-policy columnist for *CQ Weekly*, writes: "In 1968, the Iranians were among the first nations to sign the newly minted Nuclear Nonproliferation Treaty. The shah announced plans to build a network of 23 nuclear reactors for generating electricity, eight of them purchased from the United States. He was widely quoted as saying, 'Petroleum is a valuable material, much too valuable to burn. . . . We envision producing as soon as possible 23,000 megawatts of electricity using nuclear plants.' The current, less ambitious, Iranian plan is to produce 10,000 megawatts."[87]

The West countered that the real purpose of Iran's program was to develop a nuclear weapon. Before they reached a deal, Iran claimed that the West made that argument as a pretext to deprive Iran of its right to nuclear technology, argued Javad Zarif, permanent representative of the Islamic Republic of Iran, before the Security Council. He writes,

> We had a suspension for two years and on and off negotiations for three. . . . Accusing Iran of having "the intention" of acquiring nuclear weapons has, since the early 1980s, been a tool used to deprive Iran of any nuclear technology, even a light water reactor or fuel for the American-built research reactor. . . . The United States and EU3 never even took the trouble of studying various Iranian proposals: they were—from the very beginning—bent on abusing this Council and the threat of referral and sanctions as an instrument of pressure to compel Iran to abandon the exercise of its NPT guaranteed right to peaceful nuclear technology.[88]

Zarif's quotation underscores the importance of this issue to the Iranian government. Iranians perceived it as one of the main issues that unified their stance against the West.

Further, the ruling clerics' overall objective is to move Iran from being a pariah to having a position of stewardship headed by Shia Islam. Could this explain why Iran's nuclear program proves such a decisive issue among Iranians? It also explains the tensions between those who support the program and those who oppose it. Leonard Weiss, an affiliate of the Center for International Security and Cooperation at Stanford University, states, "The ruling clerics did not seek power in order to see Iran destroyed; they see themselves as stewards of a revolution that they believe will bring Shia Islam to its rightful place of world leadership."[89]

Keep in mind that the perception of the Iranian masses differs. The Iranian people would prefer to avoid confrontation with the West. Rather, they seek to join the international community. They want to improve their living standards and have access to technology, innovation, and education that goes beyond religious teachings to include the hard sciences and their benefits. Equally important, Iran's argument for not wanting to give up its nuclear program must be placed within the context of what it means for its national pride. William O. Beeman, professor of the Middle East Studies program at Brown University, who spent years in Iran, argues that the political conversation in Iranian political circles centers on the nuclear program, thus unifying political elites all of stripes. He writes, "The Iranian side of the discourse is that they want to be known and seen as a modern, developing state with a modern, developing industrial base. The history of relations between Iran and the West for the last hundred years has included Iran's developing various kinds of industrial and technological advances to prove to themselves—and to attempt to prove to the world—that they are, in fact, that kind of country."[90]

From Iran's perspective, its stance on its nuclear program goes beyond its alienated rights to nuclear technology, according to the Nuclear Nonproliferation

Treaty, of which Iran is a signatory. Rather, the nuclear program is the hallmark of its identity in defiance of the West. After all, the success of the Islamic revolution of 1979 owed neither to a genius economic strategy nor to the development of a technology that made Iranians' lives much better. Rather, the revolution succeeded because it vehemently opposed "Western hegemony" and "American imperialism." Those slogans served the religious establishment well in convincing Iranians that the West's opposition to Iran's right to nuclear technology aimed to undermine Iran's development and progress.

To counter the West's pressure regarding its nuclear program, Iran undertook a policy of exposing the double standard embraced by the UNSC and the United States. Iran argued that the standards according to the NPT, which Iran is a signatory of, were not applied to two nuclear-armed nations who failed to sign the NPT. The two countries in question are India and Pakistan, which developed nuclear-weapons capability in 1974 and 1998, respectively.

That fact cast doubt among Iranians and further cemented the notion that Iran could not trust the West to provide nuclear fuel for energy. Could that be the reason Iran vehemently objected to halting its nuclear program during the negotiating phase? There are, of course, other reasons. For example, Iran seeks regional influence, the lifting of sanctions, and further expansion of the Shia ideology. Considering the effects those aspirations may have on the region's geopolitical landscape, I argue that the potential outcome of a nuclear Iran would eventually transcend the borders of the Middle East.

Furthermore, consider South Korea's case: reports claim that it conducted undeclared nuclear experiments. Yet, the IAEA board of governors decided against filing a formal noncompliance report. Now that economic sanctions on Iran have been lifted, might Iran and South Korea begin to cooperate economically? Evidence suggests that South Korea has already begun this initiative. For instance, South Korean president Park Geun-hye visited Iran, the first visit of a sitting president in fifty-four years.[91] This cooperation is not limited only to agriculture, petroleum, and technology but also includes nuclear energy. Duyeon Kim, an expert on nuclear nonproliferation, writes, "In addition to their established economic and business ties, Iran and South Korea are also beginning to explore potential cooperation in a formerly off-limits sector: nuclear energy. Following the 2015 nuclear deal (the Joint Comprehensive Plan of Action, or JCPOA), Iran is now seeking to expand its peaceful nuclear energy ties with other countries. Meanwhile, advances made by the South Korean nuclear industry, which makes both small modular reactors and large nuclear power plants, have made it an attractive supplier in recent years."[92]

I argue that, from Iran's perspective, the nuclear program represents the ultimate opportunity to break into the international market, assuming it does not cheat; otherwise, the sanctions will be reinstated. I am also convinced that Iran will be careful how far it pushes the West when, for instance, it comes to the temptation of conducting undeclared experiments of its nuclear program.

One has reason to question whether the Iranian people are on board with the government regarding its approach to the nuclear program. The two are on opposite sides when it comes to Iran's nuclear program. For instance, Robert Tait of the *Telegraph* in London argues that there is a contradiction between what the Iranian people want and what the Iranian leadership is stating regarding the nuclear program. He writes, "The majority of Iranians want to suspend the country's nuclear program in return for a lifting of Western sanctions according to a state television poll which runs counter to claims of universal support by Iran's leaders."[93] Conducted by state-owned television, the survey sought to demonstrate that Iranians of all stripes are united behind the theocratic leadership; reality suggests otherwise.

In another blow to the Iranian government, a former general in Iran's Revolutionary Guard Corps challenged the notion that Iran's nuclear program is for peaceful purposes. Saeed Kamali Dehghan states, "A former general in Iran's Revolutionary Guards rejected the Iranian government's claims that the nuclear program is peaceful, and dismissed a fatwa issued by Ayatollah Khamenei and has accused the supreme leader, Ayatollah Ali Khamenei, of having blood on his hands over the brutal crackdown on the opposition, and described government claims that its nuclear programme is entirely peaceful as a 'sheer lie.'"[94]

I believe that Iran's leadership will continue its quest to acquire the bomb, although they will proceed carefully in order to evade detection by the West. The mechanism put forth in the agreement between Iran and the P5+1 might not be sufficient to guarantee that Iran will not one day acquire a nuclear device; one cannot be certain. Pakistan and India were able to avoid detection by Western intelligence agencies as they conducted underground nuclear bomb tests. By the time Western intelligence found out, the train had already left the station.

The bottom line: Iran has convinced itself that it is on a crusade, a race against time to achieve its objective. The only question is how far it is willing to go to get a nuclear bomb.

The Perspective of Middle Eastern Sunni Countries

Undoubtedly, Iran's nuclear agreement with the West has not only angered regional players in the Middle East but also spurred interest among them about a possible nuclear arms race in the region. Countries that expressed their concerns and yet their desire to start a nuclear program include Saudi Arabia, Egypt, Turkey, the United Arab Emirates, and, to some degree, Jordan. Notably, all the countries are Sunni Muslim. I argue that the ideological difference between Sunnis and Shiites renders the debates even more contentious. One concludes how tensions between Saudi Arabia, a Sunni country, and Iran, a Shiite country, over a host of issues, including civil wars in Syria and Yemen, led to the two Muslim rivals cutting off diplomatic relations.

Of interest in this discussion is that most Sunni countries in the Middle East, mainly key players, have expressed their support for military strikes against Iran. For instance, in 2010, the UAE's ambassador to the United States conveyed his country's support for a military strike against Iran, according to James Dorsey, a former *Wall Street Journal* foreign correspondent.[95] It is crucial, however, to explain why these countries are anxious about Iran's nuclear program. To the Sunni Muslim countries, the issue at heart is not a nuclear weapon but Iran's ability to expand its religious influence and ideological principles. While tensions between Sunnis and Shiites had lain dormant for centuries, the revival of Shiism, mainly after the US invasion of Iraq in 2003, provided Iran the perfect opportunity to expand its ideological influence. The ongoing sectarian violence in Iraq and the wars in Yemen and Syria bear out that point.

Yet, mainly, those countries fear internal uprisings of their Shiite minorities supported, or perhaps even guided, by Iran. The possibility of uprisings threatens the monarchies' hold on power. During the Arab Spring, Shia groups in Bahrain, Saudi Arabia, and Yemen challenged the government's authority. As a matter of fact, tensions persist in Bahrain, for instance, following the government crackdown on its Shiite population. As of this writing, Bahrain is still embattled by social chaos and demonstrations from its Shia majority. In this regard, Nicholas McGeehan, the Bahrain researcher at Human Rights Watch, argues that the decision to revoke Ayatollah Qassim's citizenship "takes Bahrain into the darkest days it has seen since the antigovernment protests and violent crackdown of 2011."[96]

The support, mainly financial, that some of the Sunni countries listed above are willing to provide indicates how far they are willing to go to prevent Iran from becoming a nuclear power. Those countries go so far as to describe Iran's nuclear program as an "existential threat" and "evil," according to Ian Black and Simon Tisdall of the *Guardian* magazine.[97] But tensions between Iran and other countries in the region are not new. Such tensions trace their roots to Iran's Islamic revolution of 1979 when Iran tried to export its revolutionary principles, especially to countries like Bahrain that have a sizable Shiite majority.

Since then, nations have been suspicious of Iran's foreign policies in the region. The suspicion is nowhere more prevalent than in the ongoing tensions between Sunni-dominant Saudi Arabia and Shiite-dominant Iran. The suspicion concerns not only Iran's nuclear program but also the Saudi kingdom's vehement unwillingness to relinquish its religious leadership in the Muslim world. Simon Mabon, a lecturer of international relations at Lancaster University in England, writes, "The events of 1979 in both Saudi Arabia and Iran demonstrate the importance of ideology, seen with the emergence of the Islamic Republic of Iran and the rise of Islamic opposition to Saudi Arabia. The importance of Islam within the kingdom, coupled with the emergence of a belligerent neighbor claiming legitimacy within the Islamic world, posed a serious challenge to the Al-Saud. Yet, in addition to this religious competition the two states also became increasingly involved in geopolitical competition, both in the Gulf and MENA regions."[98]

Furthermore, Iran conducts foreign policy based on terrorist activities in the hope of achieving its objective. This approach renders countries in the region, at least the ones listed above, wary of Iran's true motives in the region. Now that the West has reached an agreement with Iran over its nuclear program, it has become more compelling for key players in the region to consider their foreign policy and security options. For instance, Egypt signed an agreement with Russia in which the latter will build a nuclear power plant. The project is expected to be completed by 2022. The Egyptian president, Abdel Fattah al-Sisi, states, "This was a long dream for Egypt, to have a peaceful nuclear programme to produce electricity. This dream was there for many years and today, God willing, we are taking the first step to make it happen."[99]

Similarly, Jordan has agreed with South Korea to build a nuclear research reactor. The Korean Atomic Energy Research Institute and Daewoo Engineering and Construction have built the reactor, which will be operational in 2016.[100] Turkey, on the other hand, reached out to Japan after its failed negotiations with South Korea to launch the building of Turkey's nuclear power plant.[101] As far as Saudi Arabia goes, it has been widely reported that the kingdom has a nuclear bomb stored in Pakistan. Based on how much money Saudi Arabia provides Pakistan, one infers that, if and when the kingdom needs something in return, like a nuclear bomb, Pakistan would reciprocate.

One thing is certain: the Middle East's perception of Iran's nuclear program provides a glimpse into the future of the geopolitical landscape of the region. That future will be marked by hostilities and a nuclear arms race in a region that remains deeply sensitive about nuclear weapons. I argue that the Middle Eastern countries listed above would have to consider Israel's vehement opposition to their acquisition of nuclear technology. Simply stated, Israel wants to maintain its military superiority. The suspicion among regional players and Western powers is that Iran's nuclear ambitions are aimed at rivaling Israel.

Israel's Point of View

No country in the Middle East is more concerned about Iran's nuclear program than Israel. The Jewish state sees Iran's nuclear program as an "existential threat," and understandably so. However, statements by Israeli leaders regarding military options are not helping the cause. I believe Iran fully understands that attacking Israel is a red line. I am certain the mullahs in Tehran will think twice about embarking on such a suicidal mission. However, perception seems to hold sway over public opinion. In other words, some Israeli politicians have exaggerated the threat to the point that some within the Israeli cabinet, privately and publicly, Iran's nuclear program might present. Calls by Israeli hawks for military strikes on Iran's nuclear facilities prior to the agreement were counterproductive.

There were also political maneuverings on the part of some Israeli politicians who sought to "alert" the US intelligence community about Iran's nuclear program before the publication of the US National Intelligence Estimate. For instance, Yossi Baidatz, the head of the research division of Israeli military intelligence, was quoted as saying that Iran was "not likely" to obtain nuclear capabilities by 2010.[102] The fact remains that there is no indication that Iran would launch an attack on Israel. What we have heard from the likes of Ahmadinejad amounts to nothing but empty rhetoric for domestic consumption.

Iran fully understands the consequences of attacking Israel. It also realizes Israel's military superiority. Leonard Weiss, an affiliate of the Center for International Security and Cooperation at Stanford University, writes, "No evidence exists that the clerics ruling Iran, including the Ayatollah Khamenei, would launch a first-strike nuclear attack on Israel. Iran is aware of the Israeli capabilities for nuclear counterattack that would destroy Iran as a functioning entity for an indefinite period and wipe out significant parts of its national patrimony. Reports of the existence of a 'martyrdom movement' among Iranian women have been used to fan Israeli fears of unprovoked Iranian suicide attacks."[103]

So it behooves critics to put to rest these nonsense statements that a nuclear Iran will dominate the world, or wipe out Israel, or take over the Middle East. While I understand Iran's nuclear ambitions and growing geopolitical aspirations, reality suggests that there is a limit to Iran's aspirations. Iran has an economy that needs rebuilding, an infrastructure in desperate need of updates, and a society that remains divided over whether to continue to live within a closed system or to open up to the rest of the world and become part of the international community.

There are, however, voices of reason that see beyond Iran's nuclear threat. For instance, in April 2012, Shlomo Ben-Ami, a former Israeli foreign minister and now vice-president of the Toledo International Center for Peace, expressed his concerns about the danger of the international community failing to placate a dysfunctional region such as the Middle East. As a result, chaos would ensue, leading to far greater and more dangerous outcomes than that of a nuclear Iran.[104] The Israeli people are divided over what constitutes a threat to the Jewish state. When Israelis were asked in a poll to rank their top security concern, the majority said that Iran's nuclear program comes second to the terrorism of Hamas and Hezbollah.

Consequently, Israel is within its rights to think about its security and how a nuclear Iran will eventually alter the political calculus for Israel—and for the greater Middle East, for that matter.

The European Union's Perspective

My reference to the European Union includes countries like Germany, France, the United Kingdom, and others; however, in this brief description, my narrative focuses on key political positions reflective of Europe's perspectives on Iran's nuclear program. For instance, in June 2006, German minister of defense Franz Josef Jung suggested that Iran could use nuclear technology for civilian purposes as long as the IAEA closely monitors the program to prevent Iran from developing nuclear weapons.[105] That was then. The story now is completely different given that Iran and the P5+1 reached an agreement in the summer of 2015. Of interest in this discussion is how Germany—in addition to France, Austria, and China—rushed to Iran after sanctions were lifted in order to secure economic contracts: "This could not be more evident than in a visit by Xi Jinping to Iran; former German Chancellor Gerhard Schröder's visit to secure Germany's business interests; and a visit by a high-level French delegation to Iran for the third time since the West reached a deal with Iran over the latter's nuclear program."[106]

It is evident that Europe compromised more than the United States did on Iran's nuclear program. While there are security and geopolitical concerns on the part of the European Union toward Iran's nuclear program, the former focuses more on the economic aspects and how the lifting of sanctions on Iran brings with it unlimited economic ventures for Europe. For this reason, the European Union has never been in favor of military strikes against Iran, as evidenced by two proposals. First, French ambassador to the United States Pierre Vimont urged US policy makers to be flexible in accepting Iran's regional role while acknowledging the broad popular support among Iranians for the nuclear program. Second, the threatening language of force advocated by hawks in both Israel and the United States derailed the diplomatic process during the negotiations phase: "The threat of force had 'put some steel' into the diplomatic process," the *Economist* opines.[107] That was then. The reality now is much different given that the European Union benefits economically from the lifting of sanctions, with future economic cooperation between Iran and Europe on the horizon.

From the Europeans' perspective, Iran's nuclear agreement with the West makes sense not only geopolitically but also economically. I argue that Europe opposed any military action against Iran because the former sees Iran as an alternative energy supplier given the ongoing tensions with Russia following its annexation of Crimea. It will be interesting to see how far the European Union is willing to go to reestablish full diplomatic and economic relations with Iran. Europe's position demonstrates the disparity between one side of the Atlantic and the other regarding Iran's nuclear program.

Chinese and Russian Perspectives

I consolidate Russia's and China's perspectives into one section because, if it were not for these two countries, the Iran nuclear deal would not have happened.

Despite the Western media's reluctance to provide extensive coverage of Russia's and China's role during the negotiations phase with Iran over its nuclear program, the fact is that both Russia and China tend to benefit more from the deal than the United States or the European Union. To his credit, however, President Obama acknowledged Russia's role when he stated, "Putin and the Russian government compartmentalized on this in a way that surprised me, and we would have not achieved this agreement had it not been for Russia's willingness to stick with us and the other P5-Plus members in insisting on a strong deal."[108]

China displayed its diplomatic prowess when, in 2012, Chinese prime minister Wen Jiabao vehemently expressed his country's opposition to Iran possessing nuclear weapons. Equally important, China's warning to Iran not to close the Strait of Hormuz, as Iran threatened to do, might highlight China's growing influence and prestige on the global stage. China wants to make clear that when it comes to global affairs, those that require input from major powers, China will support other powers in condemning whatever issue is at hand. This is China's way of communicating with other global powers that it is a reliable partner and a team player.

However, China has always separated diplomatic issues from economic ones. The case of Iran's nuclear program demonstrates that distinction. Economically speaking, China realizes that it is to its benefit to support the agreement in return for lucrative contracts in the areas of energy, military hardware, and oil refineries, among others. China believes that once the sanctions on Iran are lifted, the market will be flooded with oil, thus depressing prices, which will be much better for the energy-dependent Chinese economy. So, is China the main beneficiary of Iran's nuclear agreement? The facts point to that conclusion.

From Russia's perspective, Moscow knew all along that it could hinder the efforts of the nuclear negotiations with Iran if it wanted to or if its interests, strategic and otherwise, were in jeopardy. For example, Russia knew that once sanctions on Iran were lifted, Iran would be in a better position to influence oil prices on the world market, thus hurting Russia's energy sector and putting more pressure on Russia's economy. Russia wants to kill two birds with one stone: first, when it comes to global affairs, Russia still wants to have a say and not have its input and propositions overlooked. Second, supporting an agreement over Iran's nuclear program allows Russia to invest heavily once sanctions are lifted. The Kremlin's thinking is that once the sanctions are lifted, Iran will reward Russia for its support. There is also a third scenario, which is that of China becoming Russia's patron for financing and for markets given the sanctions imposed on Russia by the United States and the European Union following the latter's annexation of Crimea. Of note: Russia has fewer and fewer options. Yet, it can still manage if it plays its economic cards right when it comes to Iran and China.

Conversely, despite competition between Iran and Russia in the energy sector, mainly in hydrocarbon production, Iran's nuclear agreement provides a great opportunity for joint cooperation between the two nations in the areas of not only energy but also armament, technology, and other sectors. One must view the dynamics among Iran, Russia, and China with the lens of a multidimensional analysis that includes energy, economic, political, and security dimensions.

Another point merits emphasis: The political debt Iran owes Russia and China must be paid not only in economic ways (through energy contracts) but also in strategic vision. The possibility exists of Iran entering into a military agreement of some sort, clearing the way for China and Russia to establish military and naval bases on Iranian soil, similar to the arrangement the United States has with Bahrain. However, it does not strike me as a wise policy. China, for instance, wants to wait till the dust settles in this latest major shift in regional geopolitics to better assess the situation and calculate its next move.[109]

The bottom line is that from China and Russia's perspective, Iran's nuclear agreement provides them a platform from which to launch, from different trajectories, their global strategic initiatives that challenge US hegemony in both the Middle East and the South China Sea. One realizes what Russia is doing in the Middle East (intervention in Syrian civil war) and Latin America (the building of a spy base in Nicaragua in a startling Cold War–like military and intelligence presence[110]), and China's posture in the South China Sea to draw the conclusion that US leadership is in decline.

Countries other than the major powers, such as Japan, India, Pakistan, and the nonaligned countries, have their own perspectives on Iran's nuclear program. While their positions differ in tone for a host of reasons, countries that support the United States tend to follow the US lead and view Iran's nuclear program as dangerous and a threat to stability. For instance, Japan's former prime minister Yoshihiko Noda expressed his country's serious concerns about Iran's nuclear program. The stance stems from, I believe, not only being supportive of the US position on the matter but also Japan's own experience when the United States dropped the atomic bomb on Hiroshima and Nagasaki in August 1945, ending World War II. However, as Iran and the West (P5+1 members) were hammering out the details before reaching a final agreement in the summer of 2015, Japan's current prime minister, Shinzo Abe, on the sidelines of a meeting of Asian and African leaders in Jakarta, expressed Japan's desire to strengthen economic ties with Iran once the agreement with the West was finalized.[111]

For its part, under the leadership of its former prime minister Manmohan Singh (2004–2014), India went from condemning Iran's nuclear program in response to US pressure to supporting close economic ties with Iran. Realizing that Iran could reach a deal with the West over its nuclear program, India rushed to reverse its position, arguing for more diplomacy to find a solution to Iran's nuclear standoff with the West. At the core of India's sudden shift is not its concern about nuclear proliferation (though India itself, clandestinely, built its nuclear capability) as it claimed. Rather, India seeks to benefit economically from a multibillion-dollar gas pipeline joint venture with Iran and Pakistan that is supposed to go into operation in 2018.[112] With the lifting of sanctions on Iran, India is well positioned to meet the demand of its growing economy and to address its energy security concerns.

Pakistan, on the other hand, has decided to pursue a policy of "neutrality" when it comes to Iran's nuclear program. As a nuclear power—one that, like India, has clandestinely built its nuclear capability—Pakistan has to be careful when deciding whether to support or oppose Iran's right to nuclear technology. Pakistan's concern stems from the argument that, when Iran develops the bomb, Pakistan will be sandwiched between two nuclear powers, India and Iran. This scenario makes Islamabad uncomfortable. Could this be why Pakistan is encouraged to build more nuclear weapons? While Pakistan could justify its stance that Iran need not develop nuclear weapons, as it did in the past, its statement would contradict the reality on the ground. Ignorantly or conveniently, Pakistan played a role when its nuclear scientist Abdul Qadeer Khan, a leading architect of Pakistan's nuclear weapons program, provided Iran with nuclear technology.[113] How convenient it is for Pakistan, a non-NPT signatory, to argue that Iran must adhere to the regulations outlined in the Nuclear Nonproliferation Treaty since Iran is a signatory, when Pakistan, itself, has not signed.

The purpose of presenting all these perspectives is to highlight the different points of view the international community holds on Iran's nuclear program. It also highlights how national and strategic interests trump any country's position on this issue.

Whether the Iran nuclear agreement is a welcome opportunity or a mistake of historic proportions, no one knows. Certainly, the agreement will bring the inevitable geopolitical shift, and its tremors will be felt beyond the borders of an ever-changing Middle East for decades to come.

5   Nuclear Arms Race in the Middle East

As of this writing, debates are under way regarding whether the Middle East is embarking on a major security and geopolitical shift, one that involves an inevitable nuclear arms race in the wake of Iran's agreement with the West over its nuclear program. Countries like Saudi Arabia, Turkey, Egypt, and Jordan have expressed a serious inclination toward acquiring nuclear technology (used herein interchangeably with the term "weapons"). In this chapter, I provide a detailed narrative about the four Middle Eastern countries listed above, which I strongly believe have serious reasons for acquiring nuclear weapons of their own rather than depending on the United States for their security and defense needs. However, the reality is that developing a nuclear program is a complex and challenging objective.

The Middle East is the most volatile region of the world and the epicenter of ongoing military, sectarian, and social conflicts. For that reason, the proliferation of nuclear weapons in the region is dangerous and destabilizing. Further, multiple countries in the Middle East with access to nuclear devices would be a nightmare scenario for the United States and Israel, leading to serious security considerations on both sides of the Atlantic. The reason is that, throughout the history of the region or the involvement of the United States there, the latter has been—and continues to be—the sole guarantor of peace and stability. Yet, never have there been challenges of such a magnitude as those the region faces today. Alas, though the international community has taken it upon itself to address the upcoming challenges of Iran's nuclear program and what might become of it in the next fifteen to twenty years, nuclear proliferation among other Middle Eastern countries has not been on the international community's agenda.

Conventional wisdom suggests that this thinking ought to change given the inevitable nuclear arms race in the Middle East. Ari Shavit writes, "An Iranian atom bomb will force Saudi Arabia, Turkey, and Egypt to acquire their own atom bombs."[1] Indeed, it has become axiomatic among Middle East watchers, nonproliferation experts, Israel's national security establishment, and a wide array of US government officials that Iranian proliferation will lead to a nuclear arms race in the Middle East. President Barack Obama himself, in a speech to the American Israel Public Affairs Committee on March 4, 2012, said that if Iran went nuclear, it was "almost certain that others in the region would feel compelled to get their own nuclear weapon."[2] That suggests that those who argue that Iran's nuclear program, and eventually the development of a nuclear bomb, will not eventually spur a nuclear arms race in the Middle East are utterly mistaken.

In this introduction, I lay out an argument for why Saudi Arabia, Turkey, Egypt, and Jordan consider the nuclear option. I claim that they seek a nuclear option for economic reasons, namely for the production of electricity to meet the demand of their growing populations, as they all claim. They also seek nuclear technology for security and geostrategic needs as Iran flexes its military, economic, and ideological muscles. I divide this chapter into four separate sections, addressing each country in the following order: Saudi Arabia, Turkey, Egypt, and Jordan. I then conclude the chapter with my perspective on what sort of decisions each country might embark on and whether it is in the security and political interests of each to pursue a nuclear weapons program that may ultimately impact the ever-changing Middle East.

Saudi Arabia

While this section focuses on the possibility of the desert kingdom pursuing a nuclear weapons program, I begin with a brief account of Saudi Arabia's place in the Middle East. According to historians and Muslim scholars, Saudi Arabia, a Middle Eastern country, is the birthplace of Islam. It lies in the heart of the

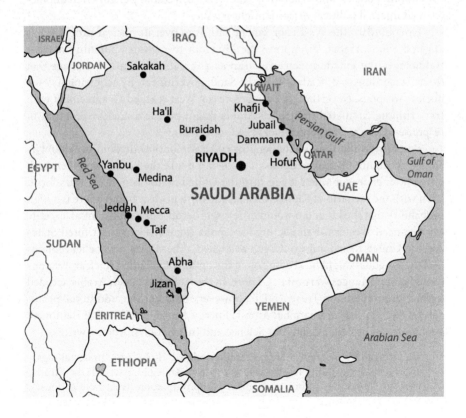

Middle East, bordering Jordan to the north, Yemen to the south, the Red Sea to the west, and the Gulf states to the east. The kingdom is also home to the two holiest shrines in the Muslim world besides Jerusalem: Mecca and Medina.

The kingdom's history is marked by an ongoing power struggle among members of the royal family. The kingdom is one of the world's top oil producers, accounting for about 17 percent of the world's oil reserves. Some estimates suggest that, in the last twenty-five years, the kingdom's oil reserves have jumped from 172.6 billion barrels in 1989 to 257.6 billion barrels in 1990, and are said to be at 267 billion barrels as of 2014.[3] Addressing Saudi Arabia's place in the world economy given its massive oil reserves or its religious influence in the Muslim world would be a great topic for another book. However, my narrative focus in this book appertains specifically to the kingdom's strategic interests and security policy given Iran's nuclear agreement with the West. Will Saudi Arabia build its own nuclear weapons program from the ground up in response to Iran's nuclear program, or will it acquire one or more nuclear bombs from Pakistan? Will Saudi Arabia seek the United States' assistance with nuclear technology, or will it turn to Russia and China? These are a few of the questions I address while exploring the options, policies, and vision that Saudi Arabia, a country engulfed in unprecedented internal political discord, might pursue.

Undoubtedly, the West's agreement with Iran over its nuclear program has angered Saudi Arabia, prompting the kingdom to consider pursuing nuclear technology. The kingdom perceives Iran as a threat and wants to manage that threat, according to officials within the Saudi government, by acquiring its own nuclear weapons. Nevertheless, even before the West reached an agreement with Iran, rumors circulated that Saudi Arabia might pursue a nuclear program in response to Iran's. Do those rumors hold some truth?

Posturing aside, that is a challenging endeavor for the kingdom, as I doubt its ability to acquire nuclear weapons of its own. I doubt the kingdom will embark on such an endeavor when it can hire the United States to fight its wars for it. Even with the scenario of Saudi Arabia acquiring a nuclear bomb, there is a high probability that the kingdom would be less safe because of its substantial appetite for advanced Western-manufactured weapons (mainly from the United States and the United Kingdom) would go unsatisfied.[4] The widely held belief remains that the kingdom might be keeping open the option of pursuing nuclear weapons should circumstances warrant it. Of note, in the late 1980s, Saudi Arabia acceded to the Nonproliferation Treaty (NPT); however, it did not sign additional protocols set forth by the International Atomic Energy Agency (IAEA). Olli Heinonen, a senior fellow at Belfer Center for Science and International Affairs, writes,

> The kingdom's current nonproliferation-related diplomatic undertakings allow it some flexibility in pursuing alternative strategies, particularly if Iran were to "break out" from its Nuclear Nonproliferation Treaty obligations. Saudi Arabia ratified the NPT in 1988 but only concluded a Comprehensive

Safeguards Agreement with the International Atomic Energy Agency (IAEA) in 2009. In doing so, it agreed to an earlier version of the "Small Quantities Protocol (SQP)" and has yet to accept the modified SQP adopted by the IAEA Board of Governors in 2005. In addition, Saudi Arabia, like Iran, has not yet signed the Additional Protocol, which allows for stricter inspections. Nor has it signed the Comprehensive Nuclear Test Ban Treaty, though it has consistently supported the establishment of a nuclear-weapons-free zone in the Middle East.[5]

Saudi Arabia's Internal Political Climate

Iran's nuclear saga, at least during the months before its agreement, occurred during an interesting time in Saudi Arabia's internal political course. In January 2015, King Salman ascended to power following the death of his half-brother, the late king Abdullah. Aware of the West's ongoing negotiations with Iran over the latter's nuclear program, King Salman wanted to reassert his authority by reorienting the kingdom's foreign-policy trajectory. Iran's growing sphere of influence proved among the key factors affecting King Salman's foreign-policy decisions—especially in the kingdom's neighboring countries of Yemen, Iraq, and Syria—which has sparked political anxiety among members of the kingdom's political upper echelon.

Further, the Saudi cabinet shake-up King Salman embarked on had more to do with demonstrating to the world his flexibility when conducting Saudi politics, a style and substance that differ from those of his predecessor and half-brother, the late king Abdullah. That said, during the West's nuclear negotiations with Iran, the kingdom surprisingly pivoted toward Russia. Saudi Arabia acts like a spoiled child who did not get his way and, instead of asking politely, then threw a tantrum (a political one, in this case). My point is that the kingdom disliked the United States negotiating with Iran over its nuclear program for fear not only that they would reach a final agreement but also that they would establish some form of rapprochement.

I cannot think of any other reason than geopolitical concerns at the heart of King Salman's political pivot toward Russia. That move offers a glimpse into King Salman's political thinking. I believe that three reasons might have led King Salman to adopt his political posture. First, the kingdom realizes the United States' declining dependence on oil resources from the Middle East—and frankly it is about time—because of US domestic oil production. Second, the shift of wealth from West to East suggests that a huge economic transformation is unfolding, offering the prospect of new partners, mainly China, South Korea, and Japan. Third, King Salman wants to send a message to the United States that Saudi Arabia can always reach out to US adversaries when the United States fails to consider Saudi Arabia's strategic interests.

Such dynamics prompt internal debates among members of the Saudi royal family over whether the kingdom should reconsider its stand on acquiring nuclear weapons. Yet, the debate has been nothing more than empty rhetoric in the hope of influencing negotiations between Iran and the West. Now that a final nuclear agreement between the West and Iran is reached, sanctions are lifted, and investments are pouring into Iran, what is the kingdom to do? King Salman's ascendancy to power brought major changes to the kingdom's approach to foreign policy. King Salman's appointment of his own son, Crown Prince Mohammed Bin Salman, to the post of defense minister (despite his youth and inexperience in global affairs) in addition to Adel al-Jubeir, Saudi Arabia's former ambassador to the United States, to the post of foreign minister, speak volumes.

Seen through a geopolitical lens, the kingdom wants to project leadership, which it lacks, and confidence, which it also lacks. Hence, it embarks on a bold, aggressive foreign policy aimed not only at countering the rise of Iran but also at pursuing international security goals that include possibly acquiring nuclear weapons. I ask: Even if Saudi Arabia decides to try to obtain nuclear weapons, which I do not foresee, does it have the enrichment technology, know-how, and necessary infrastructure to build its nuclear program? The answer is a definitive no. Saudi Arabia cannot even build a car of its own, let alone a nuclear weapon. And, even if it has some of the prerequisite industry to embark on an indigenous effort, it would most likely take decades before its nuclear program would become fully operational.

However, the kingdom has abundant funds to purchase an already-made nuclear bomb should the need arise. The question to ask is, when Shiite Iran acquires the bomb, will other countries, mainly Sunnis in the Middle East, follow suit? How might Israel remake its security and strategic posture? And how far will Saudi Arabia divert its petroleum dollars to this end as it does with the spread of its poisonous Wahhabi ideology? I pondered these questions as negotiations between the West and Iran were under way.

In the case of Saudi Arabia, conventional wisdom suggests that the desert kingdom will seek to acquire nuclear weapons from a third party; thus, an unprecedented level of proliferation will result. Ari Shavit, an Israeli reporter and author of *My Promised Land: The Triumph and Tragedy of Israel*, writes,

> An Iranian atom bomb will force Saudi Arabia, Turkey and Egypt to acquire their own atom bombs. Thus a multipolar nuclear arena will be established in the most volatile region on earth. Sooner or later, this unprecedented development will produce a nuclear event. The world we know will cease to be the world we know after Tehran, Riyadh, Cairo or Tel Aviv become the [twenty-first] century's Hiroshima. An Iranian bomb will bring about universal nuclear proliferation. Humanity's greatest achievement since 1945 was controlling nuclear armament by limiting the number of members in the exclusive nuclear club. This unfair arrangement created a world order that guaranteed relative world peace.[6]

Saudi Arabia's motive for pursuing nuclear technology differs dramatically when compared to Jordan, for instance. International security analysts argue that the fear of potential nuclear blackmail from Iran motivates countries like Saudi Arabia to pursue nuclear weapons. Could we see a scenario similar to the India-Pakistan nuclear rivalry unfold in the Middle East between Iran and Saudi Arabia? The possibility is there, but it is unlikely. We need to distinguish between the Middle East (Iran versus Saudi Arabia) and Southeast Asia (Pakistan versus India). Modern-day tribulations in the Middle East encompass a host of conflicts (the civil wars in Syria and Yemen, sectarian violence in Iraq, ISIS, failed state in Libya, security instability in Egypt, upheavals in Turkey, and ethnic tensions in Bahrain and Lebanon) that differ completely from those in Southeast Asia. These are multidimensional conflicts that include multiple players, unlike Pakistan and India, who have engaged each other militarily in four wars since 1947.

Could a similar scenario—a war, for example—erupt between Iran and Saudi Arabia? I do not think so. It would be suicide for both countries and would turn the Middle Eastern geopolitical equilibrium upside down. The region would become a multipolar system pulling world powers against the tides of history. So, let us put to rest the notion that the danger Iran presents to Saudi Arabia is similar to the one Pakistan perceives from India and vice versa; it is a meritless, nonsense argument in the debate. Even if the Saudis try to build a nuclear bomb, I do not see how they can succeed, knowing what they must overcome to achieve that goal. Only two viable options are available to Saudi Arabia when it comes to acquiring nuclear weapons: (1) taking one or a few off the shelves from Pakistan; or (2) continuing to seek protection under the US nuclear umbrella.

As far as the first option, it is widely believed that, in return for assistance when the need arises, Saudi Arabia helped fund Pakistan when it sought to acquire nuclear technology: "The Pakistanis maintain a certain number of warheads on the basis that if the Saudis ask for them they would immediately be transferred," a former Pakistani intelligence officer claimed.[7] The nuclear debate within the Saudi royal family reveals those who support acquiring the already-made nuclear bomb from Pakistan and those who suggest continued cooperation with the United States. For instance, former Saudi intelligence chief Prince Turki al-Faisal supports the idea of a clandestine cooperation with Pakistan regarding a nuclear program for military purposes. This proposition has the unwavering support of Prince Bandar Bin Sultan, the long-serving Saudi ambassador to the United States and former director of the Saudi intelligence services. On the opposite end, one finds Saudi hawks like former minister Saud al-Faisal, who prefers that the defense of the kingdom against Iran be through the United States. He argues that continuing the policy is sound since the kingdom uses American troops for its wars, betting in return for financial reward.

No matter what we hear from Saudi Arabia and the like, the discussion of nuclear proliferation amounts to empty rhetoric in the Middle East. The reason is that there is a difference between desiring nuclear weapons and having, or acquiring, the technical capability to develop a successful nuclear program. In my opinion, the only country that can acquire nuclear weapons in response to Iran is Turkey. In the subsequent pages, I detail my argument for why Turkey might forgo pursuing the nuclear option.

One might ask about Israel. I believe that, should nuclear proliferation happen in the Middle East, Israel will have to reconsider its security priorities. By doing so, Israel would make it clear to countries like Iran or Saudi Arabia that might harbor ill intentions to the point of attacking the Jewish state. I do not foresee that scenario given the grave consequences. Key players in the Middle East, including Turkey, Iran, Saudi Arabia, Egypt, and Jordan, fully understand the ramifications of Israel acting in self-defense if attacked. The United States will have to come to the aid of the Jewish state. Russia and China will intervene, and the European Union will observe from the sidelines. As a result, the Middle East will become a geopolitical ring where fighters have neither a referee nor rules to govern the fight. The bottom line is that it would be suicidal for any of these countries to attack Israel using nuclear, or even conventional, weapons.

Some Saudi officials argue that if Iran and Israel possess nuclear weapons, the kingdom will have no option but to follow suit. However, I believe that assertion is based on fear rather than rationality and pragmatism. I reiterate my previous assertion about the kingdom's lack of infrastructure, know-how, and manpower (nuclear scientists, in this case) to develop a nuclear program. Some might argue that the Saudis have plenty of money to make it happen. My answer is that, despite Saudi Arabia having access to unlimited funds, the international protocols either under the IAEA or the NPT will prevent the kingdom from embarking on a nuclear program. Saudi Arabia's unlimited funds might be useful if and when it decides to purchase an already-made nuclear bomb from Pakistan, a plausible scenario given the amount of petroleum dollars the kingdom provided the Pakistanis. But that is a long shot for one simple reason: the United States will object to such a transaction, assuming the US intelligence community is not left in the dark. Nonetheless, the kingdom's decision to embark on this venture would shift the balance of power in the region: "Saudi Arabia's reaction is a leading concern among all regional states," and "the Saudi reaction is likely to be the pivot around which inter-Arab debate resolve."[8]

The danger in such a proposition is that the internal power struggles among the Saudi royal family could precipitate a third party acquiring nuclear weapons. Even though the Saudis are under the US nuclear umbrella, I strongly believe Washington has a jaundiced view of a nuclear deal between Riyadh and Islamabad, a good enough reason for the United States to keep an eye on what the kingdom says and what it, in fact, does.

The Sunni-Shia Element in the Nuclear Debate

Another major issue ought to have the attention of global-affairs analysts, academics, and government agencies in this debate. It is the Sunni-Shia divide within the Muslim faith and how that ideological division impacts the nuclear debate not only within Saudi Arabia but also within the Muslim world at large. The current debate over whether the desert kingdom might pursue a nuclear program of its own hinges on its ability to frame the argument from a security and balance-of-power perspective, not a basic religious one. The issue in question has to do more with the schism between Sunnis (primarily in Saudi Arabia) and Shiites (dominant in Iran) for two main reasons. First, Saudi Arabia perceives Iran as a religious enemy, an ideological threat, and a dangerous political rival. This rivalry is not new, at least to those who have read the history of this conflict. Rather, it traces its roots to the aftermath of Prophet Muhammad's death in 632 AD. The main claim from the Shia perspective is that Ali, the prophet Muhammad's grandson, was the rightful successor of the Muslim *ummah*, an Arabic word for *community*. The disagreement began a tumultuous era marked by a wave of assassinations and senseless killings between the two sects of Islam that continues to mar the relationship and undermine trust between Sunnis and Shiites even today.

Second, one must appreciate the complexity of the relationship between Sunnis and Shiites to have an in-depth understanding of the ongoing rivalry between Saudi Arabia and Iran. To make matters worse, add to this equation the aspirations of both parties to acquire nuclear weapons, and the result is an eternal rift unbound by time or location, defined only by who has more nukes. The rift has extended beyond the borders of the Middle East to include countries like Pakistan, Afghanistan, India, Nigeria, and Bahrain, among others with a sizable Shia population. Within the context of a nuclear arms race in the Middle East, I wonder whether, given its past failures to understand this rift, the United States grasps the geopolitical ramifications of Saudi Arabia acquiring the nuclear bomb. I hope Washington has a strategy in place.

We cannot underestimate how much Saudi Arabia will try to justify its approach to nuclear acquisition, assuming it decides to move forward, by framing the threat Iran poses through a Sunni-Shia lens. While the kingdom argues that it fears the expansion of Shia ideology and Iran's taking over the Muslim world, the claim is preposterous given that the Shiite camp wields no influence and no power when it comes to demographics within the Muslim world. The Saudis are trying to rally other Sunni countries to support its argument. As I argue in my previous writings, the numbers do not add up to support the Saudis' claim that Iran will take over the Muslim world: "There are approximately 1.6 billion Muslims in the world representing 23 percent of the world's population. And of the 1.6 billion, only 10 percent is Shia. Even if Iran expands its influence in the Middle East, it will hardly have religious dominance over the Muslim world."[9]

The unfortunate reality is that, given these scenarios, Washington needs to coldly and objectively consider the geopolitical landscape in the Middle East at present, including the possibility of a nuclear Saudi Arabia. Alas, the Middle East geopolitical landscape is shifting and not in our favor, a good enough reason for Washington to keep an eye on a perplexing, unpredictable ally, Saudi Arabia. Indeed, it is time for Washington to assess whether it wants to tie its geostrategic interests to a confounding ally that seems not to share America's strategic vision.

In conclusion, Saudi Arabia must consider many factors when deciding whether to pursue nuclear weapons or simply rely on the United States as it has for the last seventy-plus years. While there is no one specific issue, I wonder to what extent the Saudi royal family's internal fighting affects its overall decision regarding nuclear acquisition. Given my understanding of the Arab culture, familiarity with its traditions, and fluency in the Arabic language, I feel confident in saying that Saudi Arabia will avoid the nuclear route since it knows full well its inability and incapability to develop such a complex program. The only option the desert kingdom has, if it wants to go it alone, is to acquire the bomb through Pakistan. All arguments aside, the core issue is that Saudi Arabia's decision to pursue the nuclear option will trigger other major Sunni players (Turkey, Egypt, and Jordan) in the Middle East to follow suit.

Turkey

Throughout its history, Turkey has played, and continues to play, a pivotal role in shaping global politics. With its strategic location, Turkey is considered a bridge between Europe and Asia. The Republic of Turkey is located mostly on Anatolia in Western Asia. Turkey borders eight countries: Armenia to the east; Iran to the southeast; Iraq and Syria to the south; Azerbaijan to the east; Georgia and Bulgaria to the northwest; and Greece to the west. Turkey's location between two continents renders it strategically significant. To highlight Turkey's global political significance, consider its membership in the North Atlantic Treaty Organization (NATO) since 1952 and its reputed role as one of Washington's closest allies.

Addressing the history of this fascinating country since the immigration of the Turks to the "Land of the Turks" would make an excellent topic for a future book. However, my main focus here is whether Turkey should consider pursuing nuclear technology in response to the West's final agreement with Iran over its nuclear program. Yet, it would do the readers an injustice if I were not to disclose how, following the defeat of the Ottoman Empire in World War I, Turkey was led by founder Mustafa Kamal Atatürk, who called for complete independence, which resulted in the establishment of modern-day Turkey in 1923.[10] The era of Atatürk was marked by an authoritarian style of governance, one that limited

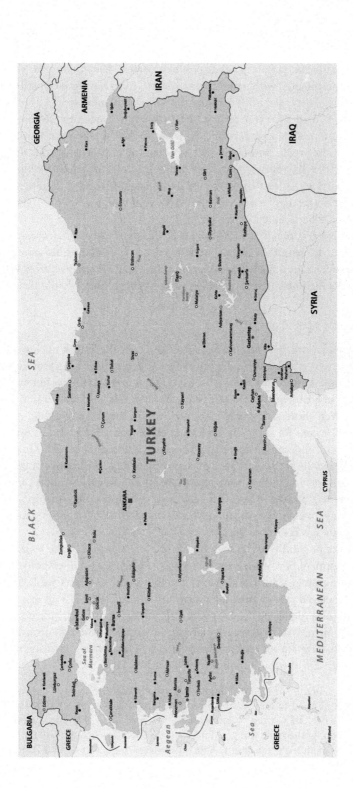

the political participation of other political parties. Fast-forward to 1950 when Turkey experimented with a multiparty system. That transition brought political turmoil, instability, and military coups in 1960, 1971, and 1980.[11]

I have to challenge the assertion that Turkey is a democratic, secular, unitary, and constitutional Islamic republic. How could that be when the country limits freedom of the press? How could that be when the president amends the constitution to serve his political ambitions while limiting the participation of other political parties? At least the so-called democracy in Turkey, despite its limitations, is far better than in other Arab and Muslim regimes. Egypt, Algeria, Saudi Arabia, and others come to mind.

All these arguments aside, the pressing question global-affairs analysts, academics, and government officials on both sides of the Atlantic are asking is this: Will Turkey pursue a nuclear weapons program in response to Iran's? What follows is a detailed narrative as to which path, I believe, Turkey will take.

The Theoretical Debate

While negotiations between the P5+1 and Iran over the latter's nuclear program were under way, there were a plethora of statements suggesting that key players in the Middle East (Saudi Arabia, Turkey, Egypt, and Jordan) would explore the possibility of acquiring their own nuclear capability in response to Iran's nuclear program. Yet, those statements amounted to nothing but absurdity. And what might be the case for that? The answer goes beyond style and substance, truth over fallacy, and pragmatism versus absurdity. Said differently, how many times have we heard the argument that a nuclear Iran would lead to proliferation in the Middle East, paving the way for Saudi Arabia (a country that imports all it needs from overseas), Turkey (a NATO member), and Egypt (a dictatorship masked as a democracy) to develop indigenous nuclear capabilities? That argument, though often repeated, is far from reality.

Yet, the possibility of Iran going nuclear is there, to say the least. But let's put the argument of a nuclear arms race in the Middle East in response to Iran's nuclear program within the context of global politics. Didn't we hear a similar argument and face comparable warnings during the Cold War era about the possibility of a global nuclear arms race? Former US officials, including Brent Scowcroft, former national security advisor to presidents Gerald Ford and George H. W. Bush, have argued this particular point. Sinan Ülgen, a visiting scholar of the Carnegie Europe think tank, writes,

> In 2009, Brent Scowcroft . . . told the Senate Foreign Relations Committee, "If Iran is allowed to go forward, in self-defense or for a variety of reasons, we could have half a dozen countries in the region and 20 or 30 more around the world doing the same thing just in case." US Secretary of State Hillary Clinton told a Senate Appropriations Subcommittee, "A nuclear armed Iran with a

deliverable weapons system is going to spark an arms race in the Middle East and the greater region." Former Bush administration official John Bolton told the United States House of Representatives' Committee on Foreign Affairs, "If Iran obtains nuclear weapons, then almost certainly Saudi Arabia will do the same, as will Egypt, Turkey and perhaps others in the region, and we risk this widespread proliferation even if it is a democratic Iran that possesses nuclear weapons."[12]

Let us assume for a moment that Turkey decides to pursue the nuclear option, which I do not foresee. Does it have the financial wherewithal and necessary infrastructure to pursue an indigenous nuclear program? Missing from debate over this issue is whether Turkey's idea of acquiring and pursuing nuclear weapons in response to Iran requires political will and committed resources. We might not know the answer. While political willingness is necessary for pursuing nuclear weapons, Turkey will also have to have the technical capability and know-how. At this stage, it is unclear whether Turkey has the much-needed resources to begin this challenging endeavor.

The other argument for why Turkey might not pursue the nuclear option is that it wants to maintain its spotless nonproliferation record. Of note: Turkey keeps a good record regarding its international nonproliferation obligations. It continues to adhere to the mechanisms and requirements set forth by the IAEA and NPT while engaging in outreach activities to promote nonproliferation in the Middle East. It will be interesting to see whether Turkey reverses its position when Iran acquires the bomb.

I support another school of thought, which suggests that Turkey is unlikely to pursue nuclear weapons even if Iran gets the nuclear bomb. As a longtime member of NATO, Turkey falls under NATO's defense umbrella, which includes nuclear deterrence. I go further, stating that Turkey's partnership with the United States should serve its defense needs. Knowing the role it plays in the United States' foreign-policy approach to both the Middle and Far East, Ankara can guarantee the military support of the United States, including the deployment of nuclear weapons should the need arise. I feel confident stating that, despite the notion that Iran represents a threat to the security and stability of the Middle East, the possibility of nuclear proliferation in response to Iran's acquiring nuclear weapons will be an insufficient reason for Turkey to try to acquire nuclear technology.

I agree with the assessments of international security analysts that suggest that Turkey must strengthen its military ties with NATO in general and the United States in particular to counter the potential threat that Iran poses. All Turkey has to do is renegotiate the terms of the presence of NATO's nuclear arsenal at Incirlik Air Base. Such negotiations could not come at a more critical time given the ongoing chaos on Turkey's border with Syria, the presence of Russian forces in the Syrian theater, and Iran's engagement in support of the al-Assad regime. Ankara, however, is not worrying too much about these developments as it expects to soon receive upgraded nuclear weapons from the United States: "If everything goes to plan, Turkey will receive the United States' newest nuclear weapon in 2019. Turkey currently hosts between 60 and 70 B61 gravity bombs at Incirlik air force base. During the Cold War, Turkish aircraft were on full nuclear alert status—meaning that Turkish aircraft were loaded with nuclear weapons and ready to take to the air in minutes, should NATO give the order. Now Turkish F-16 are only nuclear certified and would have to fly to Incirlik and pick up the bombs."[13]

As it moves forward, Turkey may face the challenge of how to convince the international community that it seeks a civilian nuclear program, not a military one. Turkey has expressed its desire to use nuclear power to meet its growing energy demands. Further, Ankara wants to rely less on Iran's and Russia's gas for electricity.[14] Certainly, one can assume that Turkey may change its mind, reversing course by pursuing a military nuclear program in the next thirty years or so. However, given its inchoate nuclear development, Turkey should not overoptimistically claim that its developing a nuclear program is feasible.

That brings me to the issue of nuclear infrastructure. Let us assume for a moment that Turkey fears an asymmetric power or the possibility of being blackmailed by Iran or others. If Turkey tries to act on such fears, its attempt to acquire nuclear capability will not go far. The reason is that Turkey lacks an infrastructure of nuclear scientists, technology, and resources. While Turkey falls in the same category as Saudi Arabia regarding its lack of infrastructure and knowhow, the Saudis have an advantage: access to unlimited petroleum dollars, allowing the desert kingdom to purchase a premade nuclear device. All things being equal, what will such dynamics mean for the next American president? The next US president will certainly face a serious geopolitical headache if the kingdom purchases a bomb: an unstable Saudi Arabia, the world's largest oil exporter, home to Islam's two holiest sites, with an abundant supply of advanced American weapons and a contingent of irate Wahhabi Sunni Muslims now armed with nuclear weapons.[15]

Turkey's Nonproliferation Policies

As stated earlier, Turkey's nonproliferation policy goes back decades. It stems from its NATO membership, which enables it to invoke Article 5 (an attack against one ally is considered as an attack against all NATO members), guaranteeing protection. Turkey also supports nonproliferation policies set forth by NPT and IAEA. Other geostrategic reasons help to explain why Turkey has embraced this policy and continues to embrace it. Chief among them are its location at the apex of the Middle East and the role it played as a frontline force during the Cold War era. As of this writing, Turkey continues to observe its international obligations to the NPT, the Chemical Weapons Convention, and the Biological Weapons Conventions; it is a signatory of them all.

To highlight its commitment to nonproliferation, Turkey reacted positively to President Obama's speech in Prague on the evening of April 4, 2009, which called for the United States to commit to nuclear disarmament. The speech followed NATO's sixtieth-anniversary summit, which President Obama attended.[16] Turkey welcomed Obama's remarks about nuclear disarmament because Ankara thinks that the president's remarks align almost perfectly with its own nuclear policy stance. Could this explain why Turkey has been advocating recently for a nuclear-weapons-free zone in the Middle East? All evidence points in that direction. It proves Ankara's strategic thinking, as one wonders, given the ongoing upheavals in the volatile Middle East, what the consequences might have been if nuclear weapons, in whatever form, were available. Alas, what is taking place on Turkey's borders makes the debate among Turkish elites more salient, emphasizing a need for stability that serves Turkey's security and long-term foreign-policy objectives.

Furthermore, although Turkey is trying to take the lead in the Middle East, its efforts have failed. Ankara's nonproliferation efforts failed to entice other countries in the region and abroad to adopt its nuclear nonproliferation policy. Sinan Ülgen writes,

> Regional upheaval and the fact that other states in the region have failed to follow Turkey's example continue to shape Ankara's nonproliferation outlook. For example, India and Pakistan are nuclear-weapon states outside of the NPT framework and have elected not to sign the treaty. Algeria, Sudan, and Israel have not signed the Biological Weapons Convention, and Egypt and Syria have thus far refused to sign the Chemical Weapons Convention. Iran is pushing ahead with its nuclear and missile programs. No state in the region is a formal member of the Missile Technology Control Regime, and many states in the region are known to have pursued nonconventional weapons in the past.[17]

Turkey's efforts to promote nonproliferation and a nuclear-free zone not only in the Middle East but also around the world gained little momentum; indeed, it is naïve of Ankara to think that major powers will give up their nukes. I foresee no scenario in which the United States and Russia will eliminate their nuclear stockpiles—nor will China, Pakistan, and India. And when the Islamic Republic of Iran acquires the bomb, there will be no incentive for it to give it up, no matter what. To Iran, obtaining nuclear weapons symbolizes its national pride, a reclamation of its identity, and, above all, a challenge to the United States-led West.

In conclusion, I remain convinced that Turkey will not get nuclear weapons in response to Iran's but rather nuclear technology to generate electricity. As a result, its policy of nuclear nonproliferation will remain constant even after Iran gets the bomb. Throughout its journey since joining NATO, Turkey faced the danger of the Cold War as a nation with perilous proximity to a nuclear-armed

USSR with its Warsaw Pact. Yet, Turkey refrained from building its own nuclear weapons back then, and I see no justification for it starting such a program today.

However, one cannot deny that a nuclear Iran stands to alter the balance of power in the Middle East. The Saudis will aggressively seek to purchase an already-made nuke (but may fail because Western intelligence will undermine those efforts), and Egypt will watch from the sidelines as it deals with internal chaos, security instability, corruption, and an angry population. These scenarios justify Turkey's strengthening ties with NATO and further ameliorating its military and security ties with the United States as the main guarantor of its defense.

Egypt

Once the epicenter of politics in the Arab and Muslim world, Egypt is now nothing but a lost cause, especially after the ouster of its long-serving president, Hosni Mubarak. But, before proceeding, it is vital to provide the reader with a snapshot of Egypt's geography and its current political landscape.

Egypt, a Middle Eastern country, is strategically located in North Africa. To the north, it borders the Mediterranean Sea; to the east, it borders Israel and the Red Sea; to the west, it shares a border with Libya, and to the south with Sudan. Throughout its history, Egypt has played an influential role in directing

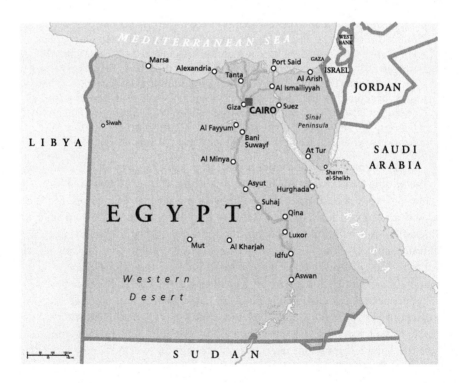

Muslim politics. For instance, Egypt was the first Muslim country to sign a treaty with Israel, in 1979, known as the Camp David Accords, when all other Muslim countries at that time refused such a treaty. The treaty led to the assassination of Egyptian president Anwar al-Sadat. As part of agreeing to a peace treaty with Israel, Egypt receives annual financial aid from the United States estimated at $1.5 billion. Only years later, under the late king Hussein, did Jordan recognize Israel and sign a treaty with it.

The political role Egypt played resulted from its long struggle to gain partial independence from Britain, which occurred in 1923. However, it was not until 1952 that Egypt acquired full sovereignty from the United Kingdom. Historians testify to the great influence Cairo has had on the Muslim political landscape. It acted not only as a bridge between the West and the Muslim world but also its spokesperson in international forums.

Economically and socially, Egypt is considered the most populated nation in the Muslim world. Of its eighty-five million inhabitants, the majority is between the ages of eighteen and thirty. Religiously speaking, Egypt is composed of 90 percent Sunni Muslims, 9 percent Coptic, and 1 percent Christians. Given Egypt's failed economic policies and corruption, the rapidly growing population has put tremendous pressure on the central government in Cairo, which is unable and unequipped to handle those challenges. To better understand how these economic pressures are pushing Egypt down the path of social revolt, one need look only at how the 2011 demonstrations in Tahrir Square in downtown Cairo reflected the population's dissatisfaction. That dissatisfaction led to the ouster of President Hosni Mubarak.

What followed was much worse, a dictator named al-Sisi, who was not democratically elected but rather ascended to power through a military coup aided by the Kingdom of Saudi Arabia and quietly approved by the United States. I address this premise in detail in the subsequent pages. For now, suffice it to say that, three years after the military coup, Egypt remains politically, socially, and economically unstable. The resulting situation has led international security experts to question whether Egypt ever will regain normalcy.

The Nuclear Debate, or Lack Thereof, within Egypt

Let us put this narrative within the context of the Middle East nuclear arms race in response to Iran's nuclear program. Egypt is one of Tehran's regional rivals. As such, it considers acquiring nuclear capabilities of its own. To what extent might Egypt be able to do so? The probability is slim to none. Evidence suggests that Egypt's announcement that it may pursue a nuclear program amounts to nothing more than empty rhetoric. How could it embark on such a highly complicated project when it has no infrastructure, no financial wherewithal, no scientific background, and no resources?

What it has instead is a corrupt political system, high unemployment, weak institutions, and a government that lacks vision, strategy, and leadership. All these factors suggest that Egypt's road to nuclear capabilities—highlighted through the statements of its dictator, al-Sisi, and his government media mouthpiece—are nothing but consumption of ink. Given the ongoing upheavals, Egypt tries to be relevant in an era dominated by jihadists' uprising in the Sinai Peninsula, ISIS, civil war in Syria, failed states in Yemen and Libya, and, above all, angry Egyptian people whose hope for democracy has been shattered beyond imagining. So, let us explore the reasons why I doubt that Egypt is capable, on its own, of developing a nuclear program for military purposes.

Vladimir Putin's visit to Egypt in 2015 sparked talks of a political, economic, and military rapprochement reminiscent of an earlier era, but I have to disagree with analysts' assessment. The reason is that Putin's objectives were (a) to secure agricultural products in response to European sanctions on Russia following its annexation of Crimea, and (b) to see whether the time was right for Russia to suggest that Egypt replace the United States as its military provider. I go further, stating that Putin's visit to Cairo gave Russia the opportunity to gauge the political mood regarding the possible expansion of its sphere of influence at a time of strong political turbulence not only in the Middle East but also for Cairo and Moscow. Egypt and Russia have also discussed the possibility of Russia building a nuclear reactor in Egypt to generate electricity for its growing population. That conversation has been at the forefront of these efforts regarding the construction of a nuclear reactor. Karl Vick writes,

> Egypt's announcement last month that it was hiring Russia to build a reactor near Alexandria made it only the latest entrant in an emerging atomic derby. Every other major Sunni power in the region has announced similar plans. And though none appear either as ambitious nor as ambiguous as what's taken place in Iran—which set out to master the entire atomic-fuel cycle, a red flag for a military program—each announcement lays down a marker in a region that, until recently, was notable as the one place on the planet where governments had made little progress on nuclear power.[18]

Middle Eastern countries like Turkey have made similar arguments to those Egypt is making, suggesting that the purpose of a nuclear program is to generate electricity. The danger is that the path to acquiring nuclear weapons begins by establishing nuclear power. Yet, in the case of Egypt, the current political, economic, and social conditions disallow forward movement in those endeavors. As of this writing, Egyptians of all stripes are gathering for demonstrations against al-Sisi's government, which promised when he took over in a military coup that he would improve conditions for all Egyptians. His promises have amounted to empty rhetoric, like the promises of most Arab leaders.

Egypt's nuclear ambitions have no basis in reality given that it lacks the necessary resources to develop a successful nuclear program. While Egypt might seek to begin a nuclear power program to generate electricity, its overall objective is to raise its profile in the Muslim world following its global political demise. Nowadays, all we hear about are daily demonstrations in Cairo, security services' brutal crackdown on citizens, the unjust incarceration of members of the press for doing their jobs, the corrupt judicial system that ignores both the rule of law and the principle of due process, and, above all, terrorist activities that include bombings in the Sinai Peninsula.

The only reason Egypt seeks nuclear power is to reacquire prestige and thereby keep up with its neighbors, including Turkey, Saudi Arabia, Jordan, and the United Arab Emirates. Egypt desperately wants to send a message to the West and to the Muslim world that it still matters when in reality it does not. Egypt desperately seeks to assert that it can speak on behalf of the Muslim world when it cannot. Egypt desperately wants to reaffirm itself as the epicenter of Muslim politics when that role ceased to exist long ago. I find it ironic that, following the military coup in Egypt, which made a mockery of democracy, Egypt still thinks it speaks for the Muslim world.

In international relations, context matters and perception matters; in the case of Egypt, the optics are not working to its favor. Nevertheless, embarking on a nuclear program will prove costly and, to put it bluntly, it may be an unrealistic ambition for the land of the pharaohs even to consider given that it is now marred by political, security, economic, and social instability. I ask, does the world want to have a country in those circumstances armed with nuclear weapons? The answer is obvious!

Implausible Propositions for Egypt

In the wake of Iran's agreement with the West over its nuclear program, many analysts on both sides of the Atlantic overstated the possibility of a Middle East nuclear arms race. They were quick to point out that the countries might embark on such an endeavor. While some aspect of their assessment holds true, they have failed to detail the reason most, if not all, the countries listed (Turkey, Saudi Arabia, Egypt, and Jordan) are in no position to acquire nuclear weapons of their own. I offer that Egypt, as I have argued concerning Turkey and Saudi Arabia before it, will not pursue the nuclear option.

Conventional wisdom suggests that Iran acquiring nuclear weapons will hasten Egypt—and other countries—to redouble their nuclear efforts. But such an action exists only in theory, not in reality, because Egypt, like Saudi Arabia, Turkey, and Jordan, lacks the necessary infrastructure and scientific knowledge to embark on such a challenging endeavor. Egypt has far more to lose if it tries. I refer to how Israel might react once it becomes aware of what the Egyptians are doing. One should not be so naïve as to think that just because there is a peace treaty between Israel and Egypt, the Jewish state will say to its citizens, "Do not worry. We are on good terms with Egypt, and nothing will impact the security of

Israel even if Egypt acquires nuclear weapons." Israeli hawks in Tel Aviv aggressively make the case that Egypt should not be allowed to acquire nuclear weapons, and rightly so. I predict that the turmoil Egypt is under will persist for some time to come. One can only imagine the resulting havoc if malcontents, radicals, and terrorist groups in Egypt get their hands on nuclear weapons; they will not think twice about using them.

In such a scenario, Israel would be compelled to either sabotage Egypt's efforts in its quest for nuclear weapons or attack Egyptian nuclear facilities like it did in Syria, using drones and US special weapons. The down side of this scenario is that the peace treaty between Israel and Egypt would be jeopardized and likely put on the shelf as tensions rise. I go further, stating that the Egyptian government, in an effort to divert its citizens' attention away from its failing policies, might encourage ill-intentioned activities across the border with Israel. This scenario must be included even in a minimum set of possibilities.

Egypt has another major issue to worry about besides Israel if it decides to pursue an indigenous nuclear weapons program. The issue in question has to do with Egypt's dependency on the United States' yearly financial aid of $1.5 billion. I argue that despite the current Egyptian government's illegitimacy, when it comes to pursuing nuclear weapons, the al-Sisi government would not want to disrupt its security arrangement with the United States, which would then limit Egypt's access to the financial support that the United States diverts to personal bank accounts in Switzerland rather than give directly to the country to pull it out of an economic slump.[19] Whether American citizens are aware of, or approve, this yearly financial assistance that has been given since 1979, when Egypt struck a peace treaty with Israel, is the subject of another discussion. For now, suffice it to say that the survival of the Egyptian regime and its security apparatus depends greatly on maintaining peace with Israel and securing financial aid from the United States. And, despite its shortsightedness and poor leadership, the Egyptian government, no matter who is in charge, will not want to risk foreign aid to pursue a nuclear weapons program of its own, the ultimate success of which is doubtful.

Egypt might imagine having its own nuclear weapons program as a way of regaining prestige, as a nostalgic reminder of its past greatness in the Arab world, but current conditions suggest otherwise. As of this writing, Egypt's tarnished image speaks volumes about its weak leadership in the Muslim world, leadership that I believe has ceased to exist. Its glory days are forgotten, thrown into the archives of history; its military, once strong and resilient, is ill equipped and psychologically demoralized. Historians, politicians, and scholars alike agree that Egypt has fallen prey to corruption and self-serving politicians who have alienated Egyptians of all stripes. Recall how former president Hosni Mubarak was grooming his son, Jamal, to take over the helm of power, as though Mubarak believed himself and his family to be bearers of the divine right of kings. Needless to say, no Arab country now speaks on behalf of the Muslim world. None has the necessary qualities to lead; none can even understand the current Arab political landscape. So, the idea that Egypt might respond to Iran's nuclear program by building its own has no merit in this debate.

Let us not forget Saudi Arabia's importance in this question. Why is Saudi Arabia important? Egypt has always been the epicenter of Arab politics. But rumors that Cairo and Riyadh might embark on nuclear programs of their own in response to Iran's will energize competition between Egypt and Saudi Arabia regarding which country leads the Muslim world. While both are Sunni Muslim countries, they do not see one another only in terms of that commonality. Egypt perceives Saudi Arabia's attempt to acquire a nuclear bomb through Pakistan or on its own as an attempt to put the nail in the coffin of Egypt's regional influence. Once again, the government in Cairo is in denial: the era of Egypt's leadership role in Muslim politics is past. I argue that the current debate in both Cairo and Riyadh now shifts from countering Iran's growing influence to competing over a leadership role in the Muslim world: a reflection of the mind set in the Arab world (short-term gains, myopia, and instant political gratification at any expense).

Iran benefits politically when Cairo and Riyadh entangle themselves in an endless row over who can claim the leadership title over the Muslim world. Both governments fail to understand that Muslims around the world do not follow titles but rather actions and deeds. Just about all Egypt and Saudi Arabia have done is offer empty rhetoric. While that might help both countries to boost their egos, it holds neither meaning nor value among the Arab masses. Egypt and Saudi Arabia both want to use their leadership position as a mask to cover their failed policies at all levels. The Egyptians want to raise nationalism as their banner; the desert kingdom of Saudi Arabia deals in Islam, knowing full well that it is the only currency with which it can purchase legitimacy in the eyes of the Muslim world.

All things being equal, Iran would be unable to play the role of a unifying force, even though it is a Muslim country. Persians are not Arabs. Thus, Arabs, be they in Cairo and Beirut or Riyadh and Algiers, would not heed the Iranians. Even if we consider the hypothetical role Iran might play, its Shia ideology would clash with that of the Sunnis, leading to bloody conflicts like those now witnessed in Syria, Yemen, Lebanon, Bahrain, Kuwait, Saudi Arabia, and Iraq. Iran understands that the strategy of expanding its sphere of influence in the Middle East requires playing the sectarian card. Thus far, its strategy is succeeding in Iraq, Yemen, Syria, Bahrain, Kuwait, and Lebanon. If either Egypt or Saudi Arabia gets the nuclear bomb in response to Iran's, it will prove more problematic for both Sunni countries as each one will try to undermine the other. I remain convinced that neither country will try to acquire nuclear weapons of their own; rather, they will both continue to depend on the United States for their security, survival, and defense needs.

To conclude this section, I reiterate that Egypt's acquiring nuclear weapons would have serious security, political, and diplomatic ramifications. The political ramifications include the possibility of Egypt destabilizing its long-secure relations with the United States and of hostilities with Israel given Egypt's proximity to the Jewish state. Israel would be on constant security alert, not knowing which group in Egypt might get its hand on the nukes.

Regarding diplomacy, I argue that, given Egypt's leadership demise in the Muslim world and current security instability in the country, competing with the de facto leader in the Muslim world, Saudi Arabia, is outside the realm of possibility. Cairo will not want to jeopardize its relations with Riyadh, since the latter provides the former with financial incentives. As a result, Egypt has no choice but to weigh the disincentives against the incentives when considering whether to purchase a nuclear weapon.

Jordan

If anything defines Jordan on the global stage, it is its long, active role in the Arab-Israeli conflict. However, before describing that role, I must provide at least a historical snapshot of this small Middle Eastern monarchy. Jordan was demarcated in the early 1920s after the Sykes-Picot Treaty. It was not until 1946 that Jordan gained its independence and subsequently became known as the Hashemite Kingdom of Jordan. Comparable in size to the US state of Indiana, Jordan borders Saudi Arabia to the south, Syria to the north, Israel and the Dead Sea to the west, and Iraq to the east. Given its vigorous, constructive involvement in the Arab-Israeli conflict during the era of the late king Hussein and subsequently his successor, current king Abdullah, Jordan managed to gain praise both from the West and from Arab or Muslim states.

However, that was not before a wave of Arab nationalism swept over the Arab world in the late 1950s and early 1960s, bringing Jordan to join Syria and Egypt in the 1967 Six-Day War against Israel. As a result of the war, Jordan lost the West Bank to Israel. An influx of Palestinian refugees continue to strain Jordan's limited resources. The year 1994 saw a dramatic change in Jordan's policy. The kingdom signed a peace treaty with Israel. To Jordan's detriment, not all segments of Jordanian society approved of the measure, leading to internal tensions within the kingdom. Following the death of King Hussein after a long battle with cancer, his eldest son, King Abdullah, assumed the throne. There is not much that one can say about the king's current political and economic performance except that the reforms he embarked on since assuming the throne are moderate, to say the least. However, the Arab Spring highlighted much deeper problems within Jordan. It saw Jordanians demonstrating against corruption and mismanagement

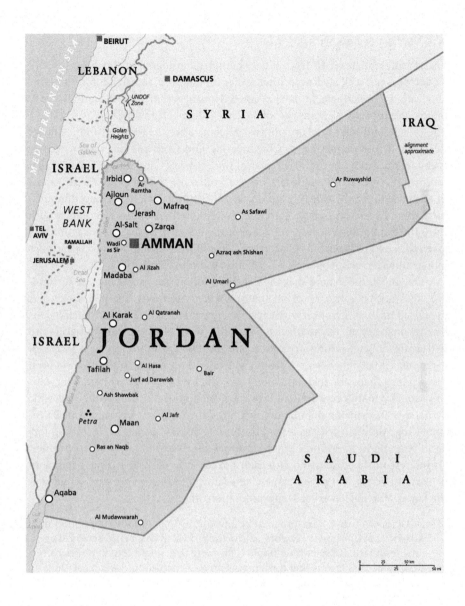

of the kingdom's limited resources. Of note: Jordan depends heavily on yearly financial assistance of $1 billion from the United States.[20]

The political role Jordan continues to play on the global stage, especially in pushing for a resolution to the Arab-Israeli conflict, would make an excellent topic for another book. However, my main focus in this narrative remains on addressing what motivates Jordan to pursue nuclear technology in response to the West's final agreement with Iran over its nuclear program.

The Nuclear Debate within Jordan

Jordan differentiates itself from other countries considered in this book because Jordan does not seek nuclear technology for military use; rather, it is for economic application. For example, Jordan has entered into negotiations with South Korea to build nuclear power plants as part of its strategy to meet growing energy demands. These negotiations resulted in a positive outcome when the Korean Atomic Energy Research Institute and Daewoo Engineering and Construction built a reactor that goes operational in 2016, the year of this writing.[21] The reactor suggests that Jordan pursues a nuclear route far different from those of Turkey and Saudi Arabia in response to Iran's nuclear program. While Saudi Arabia desires a military dimension to its nuclear pursuit, Jordan knows full well that it cannot compete with Iran. Instead, Jordan's security falls under the US security and defense umbrella given the good relationship between the two countries. This relationship includes joint counterterrorism operations and other classified projects.

One must understand that the debate in Jordan regarding nuclear technology does not take priority given that, on many occasions, the population has expressed its dissatisfaction with the government's failed economic policies. Unlike its neighbor to the south, Saudi Arabia, which has unrestricted access to petroleum dollars, Jordan's resources are limited. Quite frankly, if it were not for annual financial assistance from the United States, Jordan would have been thrown into chaos and instability long ago. The Jordanian government's forward thinking, no matter how limited it may be, concerns how to meet the economic demand of its growing population, including the influx of refugees from Palestine. King Abdullah tries to attend to those legitimate concerns, including Jordan's ongoing importation of more than 96 percent of the power it consumes. Of interest in this discussion is Jordan's decision to seek Russia's assistance in constructing a nuclear power plant. Chen Kane, Middle East projects manager at the James Martin Center for Nonproliferation Studies, writes,

> In October of this year, Jordan announced it had chosen Russia to build its first two nuclear-power reactors. Historically, Jordan has lacked access to energy resources. It depends on imports for more than 96 percent of power consumption. This means that a whopping 20 to 25 percent of Jordan's national expenditures go to importing energy. That is a massive outflow of capital for a country of only 6.5 million people. Jordan's decision to turn to nuclear power, however, does not mean that the kingdom is about to sail smoothly into the club of nations that produce their own nuclear energy. While Jordan is in great need of a less costly and more reliable energy source, it won't get there unless it can overcome some major challenges.[22]

As of this writing, the ongoing debate within Jordan is that its acquisition of nuclear technology aims strictly to expand and diversify its energy sources since the Jordanian government anticipates electricity consumption to double by 2030. I argue that Jordan's investments in nuclear infrastructure make sense given its limited resources. Jordan needs to invest in nuclear energy not only to provide electricity for households, water desalinization plants, and industry but also to free the government from some of its monetary constraints so that it can invest in other sectors of the economy.

Equally important, Jordan is rich in uranium reserves. As such, the country can export this raw material, providing it with new revenue sources. This comes on the heels of other regional countries (Turkey, Saudi Arabia, Egypt, and the UAE) announcing their intention to pursue nuclear energy. Said differently, Jordan does not want to be left out when other countries, mainly its neighbors, use and benefit from nuclear energy. Simultaneously, Jordan wants to capitalize on its uranium reserves by becoming a leading uranium provider.

The Inevitable Challenges Ahead

On the one hand, one could say that Jordan's nuclear ambitions will face major challenges. It is essential to reiterate the kind of nuclear program Jordan wants to pursue. If we are talking about a nuclear program with military application, then it is not only challenging for the Hashemite kingdom to do but also unrealistic even to consider. Like its neighbor to the south, Saudi Arabia, Jordan lacks the ready infrastructure, scientific knowledge, and manpower (scientists, in this case). Similarly, if Jordan's nuclear ambitions are strictly for economic purposes, it can still face major challenges. Chief among them is that the kingdom has no experience in operating nuclear reactors or fuel and waste facilities.

On the other hand, the possibility that Jordanians of all stripes would oppose their country's nuclear ambitions casts a shadow over the country's ability to develop a successful nuclear program. Of interest in this discussion is how Jordanians express their vehement opposition to the presence of nuclear power plants for fear that a disastrous accident like the one at Japan's Fukushima Daiichi Nuclear Power Station in 2011 might occur. Additionally, the Jordanian government might be surprised to learn of the environmental concerns that the presence of nuclear power plants in the country could have on air quality, for instance.

The Jordanian government finds itself immersed in another challenge, namely whether it has been fully transparent with the Jordanian people regarding both the cost and economic practicality of developing a nuclear program. Critics ask how the government could pay for such an expensive venture when it depends on foreign aid from countries like Saudi Arabia and the United States. The question is legitimate, particularly given how data can be manipulated to further a particular agenda and justify a particular cause. In the case of Jordan's nuclear program, Jordanians are within their rights to challenge the government's statements on this question. Ayoub Abu Dayyeh, head of the prominent Jordanian nonprofit organization the Society of Energy Saving and Sustainable Environment, argues, "In Jordan we have witnessed fraudulent elections, a fraudulent Parliament; it is not out of the realm of possibility that at the end of the day we will receive fraudulent studies."[23]

Jordan's challenges regarding the nuclear program are not limited only to the financial and scientific aspects. Rather, they extend to other areas, including geography. Given its limited acreage, Jordan finds it challenging to identify a site on which to build its reactors. That is especially problematic given that the reactors have to be within close proximity to areas with access to cooling water. Unfortunately, local tribes control those areas rather than the government. One sees the dilemma.

Would it be fair to say that Jordan is at a crossroads when it comes to the hype about its nuclear program? The answer is not as simple as one thinks. On the one hand, realities that include lack of financial resources, infrastructure, and scientific knowledge prevent Jordan from developing a nuclear program. On the other hand, given its sagging economy, high unemployment, and ongoing flow of refugees from both Palestine and now Syria, developing a nuclear program will be an expensive, unrealistic ambition. I argue that the danger lies in Jordan deciding to develop its nuclear program with limited access to funds; thus, the government might be compelled to compromise, cutting corners on environmental safety regulations, mandatory inspections, or both. Nevertheless, I question how realistic it is for Jordan even to *entertain* the idea of embarking on a nuclear program in response to Iran's. Yet, I wanted to present this narrative to highlight the possible outcome, political instability, public safety, and economic outcome that Jordan might face.

If Jordan's nuclear energy program is successful, a long shot by any measure, it could alleviate some of the economic and energy pressures the country continues to endure. However, if the nuclear program has military dimensions, it could destabilize the monarchy and, given its proximity to Syria, Iran, and Egypt, certainly would have major geopolitical implications that could reverberate well beyond the borders of the Middle East.

To conclude this chapter and put the narrative herein within the geopolitical context, I argue that the Middle East's nuclear domino theory might be just that—a theory. Statements Saudi Arabia, Turkey, and Egypt have issued on pursuing an indigenous nuclear program amount to empty rhetoric, as reality and factual evidence suggest otherwise. Yet, the possibility that the countries in question *could* impact the geopolitical landscape in the Middle East runs contrary to their capabilities of guaranteeing a successful nuclear program. Alas, throughout history, Arab nations have done more talking than doing, and the pursuit of a nuclear weapons program should be no different.

6 The Future of the Middle East and Its Geopolitical Outcome

It is a challenging proposition, an audacious undertaking to forecast what the world might look like in the years following Iran's nuclear agreement with the West. I find no point of reference, no images, and no precedents on which to base my predication. However, I *can* create a possible scenario that addresses potential issues inherent in the upcoming geopolitical shift in the Middle East. When forecasting the scope of this shift, I consider the vivid imagery of human experiences, including mine; political conditions; and historical events, both distant and recent, as I ask myself, could my prediction become a reality?

Analysts from all disciplines—military, financial, security, and counterterrorism—are wondering what will become of the Middle East once Iran acquires the nuclear bomb. What will the region's geopolitical landscape look like? As of

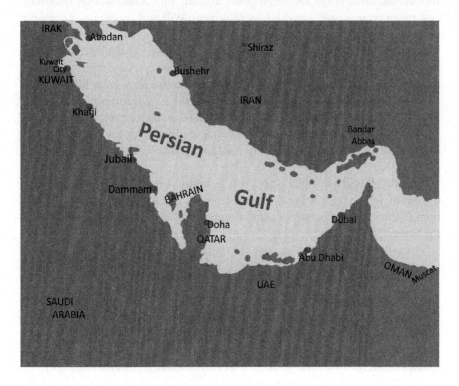

this writing, so-called policy experts debate these questions behind the scenes not only to raise awareness of the potential danger of Iran acquiring a nuclear bomb but also to talk through what the United States needs to do to manage this inevitable geopolitical shift. Yet, the US foreign-policy machinery of past and present administrations produces only the vaguest, most imprecise of strategic responses. The limited political debate within the US foreign-policy establishment centers on the concept of long-term containment. The question, however, is whether this strategy will be similar to that the United States embarked on against the former USSR or a far more complicated process that defies the norms and boundaries of international relations. The only certainty in all of this is that, one way or another, a nuclear Iran will contribute to a seismic geopolitical shift with movements reverberating far beyond the borders of the Middle East.

In the following narrative, I explore whether a nuclear Iran would be a challenge for the United States to contain. Further, I assess to what degree a nuclear Iran at the heart of the Middle East might lead major powers to engage in a cold war that defines and determines each country's geostrategic interests. Note that such interests would be shaped and influenced by both the behavior and the long-term political aspirations of a nuclear Iran.

I divide this chapter into three sections. The first section addresses the constitutive elements of the future geopolitical shift. The second section questions how major powers approach this inevitable shift: Will they cooperate or compete against each other for influence and access to the Middle East? The third section assesses whether the United States has a strategy in place to deal with a nuclear Iran and how that might shape American foreign policy in the region. Of note: each strategy carries risks that could have major consequences for all actors on the global stage. Further, this section assesses the new realities of the Middle East should Iran acquire the bomb. Will this new reality complicate matters for Washington and contribute to its ambiguous foreign policy toward the Middle East? Or will the United States realize that living with a nuclear Iran is the new order de jour?

Finally, this chapter places its analysis within the context of reports coming out of Tehran suggesting that some Iranians cite unimproved living conditions to show their ambivalence about the nuclear deal Iran reached recently with the West.

The Dimensions of the Middle East Geopolitical Shift

Debate is already under way regarding how the geopolitical shift should unfold in the Middle East in response to Iran acquiring nuclear weapons. Global-affairs analysts and academics agree that the geopolitical shift is inevitable, and eventually the United States will have no choice but to refocus its efforts on how it intends to address, and deal with, an Iran with nuclear first-strike capability.

It would be shortsighted to discuss only the political or economic aspects of the change in the geopolitical landscape of the Middle East. A nuclear Iran is a profound happening, one that would revolutionize the political thinking among Middle Eastern countries and draw in great powers, whether they like it or not. Imagine a new political order in the Middle East in which Iran politically, economically, ideologically, and religiously controls one-half of the region, while the Kingdom of Saudi Arabia controls the other half.

I agree wholeheartedly with critics who argue that such development is already under way. However, my argument is that the geopolitical shift associated with a postnuclear Iran goes beyond empty rhetoric. It includes the strategic reorganization of the national assets of all involved countries. The new geopolitical landscape postnuclear Iran will force countries like Saudi Arabia and Turkey to reposition themselves to limit the impact of this shift. For instance, Turkey's interest in the outcome of the conflict in Syria is no small matter. Its interest in the northern part of both Syria and Iraq has nothing to do with supporting minority Sunnis or ensuring the Kurdistan Workers' Party does not expand and gain influence. Rather, Ankara's strategic vision seeks to extend Turkey's influence beyond Anatolia. Reva Goujan, a security analyst at Stratfor, writes,

> We then take a long look out into the future. Turkey's interest in northern Syria and northern Iraq is not an abstraction triggered by a group of religious fanatics calling themselves the Islamic State; it is the bypass, intersection and reinforcement of multiple geopolitical wavelengths creating an invisible force behind Ankara to re-extend Turkey's formal and informal boundaries beyond Anatolia. To understand just how far Turkey extends and at what point it inevitably contracts again, we must examine the intersecting wavelengths emanating from Baghdad, Damascus, Moscow, Washington, Arbil and Riyadh. As long as Syria is engulfed in civil war, its wavelength will be too weak to interfere with Turkey's ambitions for northern Syria, but a rehabilitated Iran could interfere through Kurdistan and block Turkey farther to the east.[1]

Similarly, Iran most likely has developed plans to counter any strategic move by either its archenemy, Saudi Arabia, or Turkey.

Furthermore, Iran will likely target the Kingdom of Saudi Arabia more so than Turkey. Oil is the reason. While both Saudi Arabia and Iran are engaged in a religious cold war, Iran's concerns are more politically and economically oriented than religiously driven. Given that sanctions have been lifted, Iran wants to sell its oil on the open market while being well compensated. Of note: Iran produces around 3.4 million barrels of oil per day, according to US energy information, thus rendering Iran the world's sixth-largest oil producer.[2] Once Iran addresses its economics, its next area of concentration is its military. In other words, the geopolitical shift in the Middle East will certainly include (a) military dimensions and (b) diplomatic elements.

On the military front, I argue that once Iran secures economic access to global weapons markets and acquires nuclear weapons with delivery systems, including advanced weapons systems from Russia and China, it will play a far greater and more influential role in the region than it already does. For instance, Iran has received the delivery of Russia's S-300 air-defense systems with the additional launchers expected by the end of 2016.[3] Similarly, Iran has already reached out to China to improve bilateral relations that include military cooperation. This relationship serves both countries' economic, military, and strategic interests. Joel Wuthnow, a research fellow at the Center for the Study of Chinese Military Affairs, writes,

> In the coming years, China might attempt to sell a wide variety of advanced arms to Iran. These could include J-10 fighters, the possible sale of which has been reported in Chinese media. Another system would be the *Houbei*-class high-speed missile boat, which China is also poised to sell to Pakistan. This would be a logical choice given the expanding navy-to-navy relationship between China and Iran. China might also transfer advanced cruise missiles and technical know-how, allowing Iran to improve its domestic cruise missile program. Other systems could include UAVs, space and counter-space weapons, missile defense components and electronic warfare systems. Sales of most of these items would likely require UN approval, which China might attempt to secure as a permanent member of the Security Council.[4]

Further, major global players, mainly China and Russia, will not miss the opportunity to shape, or at least play a pivotal role in shaping, the geopolitical landscape, especially when their interests are at stake. Consider, for example that China and Russia benefit not only from the lifting of sanctions on Iran but also from the ongoing chaos in Syria, Egypt, and Yemen. Furthermore, both nations benefit from the diplomatic row between Iran and Saudi Arabia. Those upheavals allow them to maneuver politically, steering their policies to challenge the United States and oppose any policy that might jeopardize their interests, including the expansion of their military outreach in the Middle East.

As I argued in my previous writings, it is probable that either China or Russia will enter into a strategic partnership with Iran that will eventually develop into some sort of military alliance. Could we one day see either Russia or China establish a naval base in Iran? The possibility is there more so for China than for Russia. Would that arrangement be similar to the one the United States has with Bahrain? The ongoing chaos in the region makes it hard to tell at this point. I predict that China will wait until the dust settles on Syria's outcome, Libya's turmoil, Yemen's civil war, Iran and Saudi Arabia's diplomatic row, and now the latest failed military coup attempt in Turkey. China strikes me as a pragmatic state that prefers to make a strategic decision only after the ongoing chaos in the Middle East subsides. Doing so allows Beijing to assess whether it is time to challenge the United States in the Middle East through an assertive foreign policy or maintain its neutrality while reaping benefits as the United States gets involved, in blood and treasure, in endless regional conflicts. China's acquisition of major oil contracts in Iraq exemplifies its pragmatism.

As Iran expands its military apparatus by acquiring advanced military weapons from Russia and China, regional powers like Saudi Arabia and Turkey will be forced to reconsider their military readiness. I take a special interest in the case of Saudi Arabia given that it continues to be engaged in Yemen, Syria, Afghanistan, and pockets in Southeast Asia that include Bangladesh, Pakistan, and Kashmir. These tensions have far-reaching consequences that lie beyond military or economic matters. These tensions involve ideological and religious dimensions since they threaten the survival of the Saudi monarchy. Yet, at the core of this threat to the monarchy is not Iran, as Riyadh claims, but rather internal fighting among royal family members over succession.

When ugly royal infighting spills over to the outside world, international security and global-affairs analysts are going to ask the inevitable question: Is it the beginning of the end for the Kingdom of Saudi Arabia as we know it? Even mere questions like that can have reverberations that are not only economic and political but also religious and geopolitical. Given the above, I argue that the threat to the Saudi monarchy is more internal than external. While I agree that Iran's geopolitical aspirations in the Middle East are real and present a threat, they do not come close to what an internal coup from within the kingdom, for instance, might lead to. Consider my previous observations:

> Where from here? Russia's airstrikes in Syria underscore a broader threat to the kingdom: Put all the problems together and Saudi Arabia, more than ever, looks politically vulnerable. Its dependence on the United States for its survival the last 70 years seems to be near an end. The United States is no longer in position to play its traditional role as the only guarantor of Middle East stability. One can only imagine the scenario in which the house of Al-Saud is forced to relinquish power to another entity from within that does not share Washington's aspirations and/or agenda. That means our next president will face one more serious geopolitical headache: an unstable Saudi Arabia, the world's largest exporter of oil, site of Islam's holiest sites and a country equally bountiful in advanced American weapons and very angry Wahhabi Sunni Muslims.[5]

And, if this is not enough for the jittery kingdom, consider the significance of the July fourth attacks in Medina that ISIS conducted. The attacks prove that the kingdom is not only tasting its own medicine but also cannot control those same elements it built to spread its poisonous Wahhabism ideology—an ideology that has twisted the true meaning of Islam through clever commentaries and misinterpretation. One must understand that Saudi Arabia's spread of Wahhabism from Pakistan and Afghanistan to Somalia and Sudan has been facilitated by the abundance of the kingdom's petrodollars. Gregory Gause III, professor of international affairs at the Bush School of Government and Public Service of Texas

A&M University, writes, "It is undoubtedly true that Saudi Arabia, since the 1960s, has built a set of institutions and networks to spread its puritanical, narrow-minded views on appropriate religious practice, as well as intolerance of other religions or other interpretations of Islam. With the oil revolution of the 1970s, the Saudis had enormous resources to support that effort. In the 1980s, the Saudis (along with the United States) supported a campaign in Afghanistan against the Soviet Union that both they and Washington were happy to call a jihad."[6]

Saudi Arabia wants the United States to continue to do its war bidding for it against Iran in Syria, Yemen, Iraq, Afghanistan, and any other country that represents a real or imagined threat to its survival. It is just a matter of time before Saudi Arabia has to face the radical groups that it creates, funds, and supports. It will be interesting to see what sort of strategy the kingdom undertakes to deal with this inevitable challenge. Will Saudi Arabia enter into an agreement—a truce of some sort—in which it will not pursue the terrorist group(s) as along as the latter does not conduct operations inside the kingdom? That is a possibility.

To put all of this within the geopolitical context, I am certain that Iran has already considered this scenario. Should it become a reality, count on Iran supporting the Shia minority inside the kingdom's southeastern region of ash-Sharqiyyah. Ongoing fighting in Yemen between the Houthis, Shia's rebels supported by Iran, and the Saudi forces shows one example of such support.

The issue is even more complicated after adding Russia and China to the mix. How so? The answer lies in the ongoing upheavals that provide both China and Russia a platform from which to launch their strategic visions: challenging the United States and exerting their influence in an ever-changing Middle East. Because of those strategic visions, it is just a matter of time before China and Russia enter into strategic alliances with Iran, military ones to be specific, to serve their separate objectives, whatever those may be. That leads me to ask the following question: Will Russia's or China's assertive foreign policy in the Middle East collide with that of the United States? While that scenario is possible, I lean more toward seeing this possibility become a reality in the Pacific rather than the Middle East. Yet, the United States' leadership decline and disastrous foreign policies in the last fifteen-odd years send a strong message to China and Russia to be even more aggressive as the United States struggles through this ambiguous time in its foreign policy. The civil war in Syria and the South China Sea tensions strengthen my argument.

With this inevitable shift in the geopolitical landscape of the Middle East, I am more convinced than ever that, despite the separate agendas of the triangle countries (Russia, China, and Iran), they will coordinate their efforts toward the common goal of undermining the United States' efforts in the Middle East. And this cooperation has already been implemented; look no further than Syria. All things being equal, the ongoing geopolitical shift will continue to see China, for instance, moving forward with its quiet foreign-policy strategy, one based on neutrality, for the time being, because it makes sense for Beijing to let the United States carry the load and sacrifice blood and treasure while China reaps the benefits when the right opportunities present themselves. In Russia's case, I believe that Moscow realizes the United States' inability to act decisively when it matters the most, for instance, regarding the civil war in Syria, the chaos in Yemen, and the failed state of Libya. As a result, Russia will continue its assertive foreign policy in the Middle East and, while it's at it, also test the waters to see how far it can push the North Atlantic Treaty Alliance (NATO).

As far as Iran goes, Tehran undoubtedly plays a pivotal role in this geopolitical shift for two main reasons. First, Tehran fully understands that it is now or never to expand and pursue its geopolitical aspirations in the region: taking over the political process in Iraq, exploring Iraq's southern oil fields, influencing the outcome of the Syrian civil war to its favor, ensuring victory of the Houthi rebels in Yemen, and certainly establishing other Shia militias of the likes of Hezbollah. Second, the nuclear agreement with the West provides the ultimate opportunity for Iran to communicate to the world that it has challenged the United States—and the West, for that matter—and won concessions concerning its nuclear program.

Speaking of Iran's nuclear program, a recent report from the German ministry of interior reveals that Iran is engaged in aggressive efforts to acquire nuclear technology through a third party. Countries that could help Iran include China, Turkey, and, yes, one of our Gulf allies, the United Arab Emirates.[7] Iran's purported efforts support my argument that Iran seeks to get nuclear weapons and that it is just a matter of time before the Islamic republic succeeds.

I see the upcoming geopolitical shift in the Middle East after Iran's acquisition of nuclear weapons as a volcanic eruption when compared to other political shifts elsewhere—in Latin America or Africa, for example. The reason is that the Middle East is a highly volatile region and will remain so. I do not see how major powers could ever reach an agreement to share power and thereby equally benefit from the natural resources while Iran is building its nuclear capability. In support of my argument, one does not have to look far to realize how Turkey, for instance, is departing from the timid foreign policy that observers have been accustomed to. The failed military coup in Turkey on July 15, 2016, certainly provides an opportunity for President Recep Tayyip Erdogan to crack down even harder on the opposition, thus consolidating more power. These dynamics challenge the opinion held by political elites in the United States and elsewhere that Turkey is capable of challenging its competitors and foes alike, including Iran and Russia. Make no mistake: Turkey is a major player that could tilt the balance of power in either direction. Stated differently, Turkey can help to further the West's strategic interests as long as Turkey's own interests are taken into consideration. However, I can see a scenario in which Ankara goes against the aspirations of its Western allies. The West needs to understand that Turkey's geopolitical aspirations differ in style and substance from those of its allies, including the United States.

What about the Muslim world, mainly those in the Middle East? How do they see this geopolitical shift? The answer varies from one Muslim country to the next. Some would support both Iran's goal of gaining influence in the region and its nuclear quest. Of interest are countries in the Gulf. For example, could this geopolitical shift convince those in Kuwait, Qatar, and Bahrain, given their Shia demographics, to follow Iran's newfound regional preeminence? I argue, as Wall Street journalist Bret Stephens did, that Iran might eventually be in position to convince these Gulf countries to start downgrading their military ties with the United States or shut down US bases entirely.[8] Mind you, Iran could also learn from its past mistakes, especially during the US invasion of Iraq. Through its proxies and militias, Iran engaged US forces in violent armed encounters throughout Iraq, which eventually triggered a public backlash. I am convinced that Iranian elites in Tehran would avoid repeating the same mistakes.

Another variable in the equation of geopolitical shift in the Middle East is the ongoing tension that Iran continues to endure with its archenemy, Saudi Arabia. The possibility of Saudi Arabia acquiring nuclear weapons of its own in response to Iran is slim to none; that said, the ongoing tensions would have significant effects at all levels (economical, political, and ideological). Could this scenario force the United States to withdraw from the Middle East, as some have suggested? I doubt it. The United States and its Middle Eastern allies do not want to see China or Russia—or Iran, for that matter—become dominant forces in the region. Yet, in the case of Iran, it might be a forgone conclusion.

However, Riyadh fears most the possibility of a rapprochement between the United States and Iran given the latter's nuclear agreement.[9] I believe that Washington is carefully evaluating the ongoing tensions between the two Muslim rivals (Saudi Arabia and Iran) for fear that history might repeat itself. I refer to the two countries' severance of diplomatic ties from 1988 to 1991. I need not remind readers that a diplomatic rupture between Iran (Shia dominant) and Saudi Arabia (Sunni dominant) will spill over, if it has not already, across not only the Middle East but also throughout the Muslim world. The civil war in Syria and Yemen and tensions in Pakistan and Afghanistan show such spillovers already happening.

The ongoing diplomatic row between Saudi Arabia and Iran reflects their rivalry over influence in the Middle East. On the one hand, Iran considers its agreement with the West over its nuclear program a pretext for expanding its influence, though Iran's geopolitical aspirations manifested only a few years ago. On the other hand, Saudi Arabia fears losing its religious leadership and control in the Muslim world since it can deal in that currency alone when it comes to global governance or stewardship. However, I challenge critics who argue that Iran wants to take over the Muslim world. Iran has no interest in that because it realizes that its minority Shia population of about 170 million worldwide cannot control a Sunni population of roughly 1.5 billion.

One ought to evaluate from a realistic perspective what the upcoming geopolitical shift might mean for American interests, international relations, the interaction of global powers, and a volatile, ever-changing Middle East. The pages of history testify that a nuclear Iran's impact on the geopolitical landscape in the Middle East is a forgone conclusion, whether we like it or not.

The subsequent section focuses on what this geopolitical shift in the Middle East postnuclear Iran might mean for relations among global powers and how they might adjust their geostrategic calculations to ensure the security of their strategic interests, whatever they may be.

Major Powers' Approach to the New Geopolitical Landscape

In my opinion, in the wake of its nuclear deal with Iran, the United States needs to be concerned about the joint China-Russia alliance more than anything. Why is that alliance the number-one concern? The answer lies in the inevitable geopolitical shift in the Middle East. Critics argue that the alliance seeks primarily to undermine US hegemony in the Pacific, which is true. However, I have a different take on the long-term strategy of that alliance and its likely effect on US foreign policy in general in the upcoming years. Although I agree that China and Russia prefer to limit and reduce US presence in Asia, both Beijing and Moscow realized throughout the negotiations phase of Iran's nuclear program that the outcome would provide them a great opportunity to develop their plans to undermine US leadership.

China seeks to undermine US influence in the Middle East in order to ultimately divert America's focus from the Pacific. China has adamantly claimed that the South China Sea islands are part of its territory and it will not give them up. As of this writing, China has refused to accept the ruling of the International Criminal Court regarding the disputed islands. The tribunal argument is that China has violated the Philippines' sovereign rights in its exclusive economic zone by interfering with Philippine fishing and petroleum explorations.[10] Is this the beginning of a military conflict between China and its neighbors? Hardly. Will the United States get involved militarily? Possibly, given its existing treaty with the Philippines. Article 5 of the treaty states, "For the purpose of Article IV, an armed attack on either of the Parties is deemed to include an armed attack on the metropolitan territory of either of the Parties, or on the island territories under its jurisdiction in the Pacific or on its armed forces, public vessels or aircraft in the Pacific."[11] That clause binds the United States to its defense commitment.

However, I am certain that Washington will try to deescalate, knowing that China refuses to give ground in any context. To China, backing down would damage its global image, status, and prestige. Realistically, I do not see how the United States and China will engage each other militarily because both nations have too much to lose if they do. A military confrontation between the two powers is unlikely for two reasons. First, China strikes me as pragmatic in not wanting to jeopardize its steady global economic, military, and financial emergence. In doing so, China would risk creating havoc in a world system that has, after all, allowed it to transition from an agricultural society to a global economic powerhouse, a system in which China has acquired its new wealth, power, political influence, and, yes, status. Second, China understands that its military capabilities, especially its naval assets, are no match for those of the United States. For that reason, Beijing is now working hard to improve its nuclear deterrence, naval apparatus, and missile technology. Nonetheless, China is making clear to Washington that, at least in the Pacific realm, the United States' leadership days are numbered, as the territorial challenge suggests.

From China's perspective, it makes sense to expand its presence in the Middle East, and the perfect conduit to do so is through none other than Iran's newly negotiated nuclear agreement. For this reason, I believe that China will base its foreign policy in the coming years, be it in the Middle East or the Pacific, on strategic decisions rather than short-term gains, a good reason to continue its pragmatic military disengagement of the United States, this time around. Make no mistake: if China's interests are threatened to the point of, for instance, choking its economy or limiting access of its products to other global markets, it will consider the military option as a last resort. And a military confrontation with China is no small matter. Could this explain why China made the decision to cooperate with Russia? Recently, China purchased some advanced Russian weapons systems as part of modernizing its military. Yet, the cooperation has, in my opinion, other long-term objectives: "China's military cooperation with Russia is growing stronger. It includes purchase of a new Russian submarine design called the 'Amur 1650.' This cooperation serves both countries' interests. On one hand, China asserts its regional influence while modernizing its military force to deter U.S. meddling in Asia. Russia, on the other hand, partially avoids economic collapse through the sale of advanced submarine designs to China. Other deals between the two loom."[12]

While we're at it, it behooves us to evaluate these dynamics from Russia's perspective. I argue that Russia's political calculations have different objectives from those of China—or Iran, for that matter. From Iran's perspective, the shifting geopolitical landscape consists of deepening the rift between Russia and the West following the latter's ongoing sanctions on Russia in the aftermath of its annexation of Crimea. Iran sees the opportunity to keep Russia at odds with the West. This way it can benefit economically and militarily from Russia's cooperation.

Similarly, Russia is also interested in improving relations with Iran, knowing the role Iran plays in the geopolitical shift in the Middle East following its agreement with the West over its nuclear program. I will not be surprised when Russia provides additional support to Iran in the nuclear technology realm; whether this support is for civilian or military applications is the subject of another discussion.

For now, suffice it to say that both Iran and Russia realize that their interests converge on similar objectives, yet from different trajectories. This is evidenced by the Russian defense minister Sergei Shoigu's visit to Iran, suggesting that the two powers are stepping up their military cooperation.[13] My assumption is that a visit at that level suggests that something strategic is in the works between Moscow and Tehran. This rapprochement serves both countries' interests in the Middle East though from two separate outlooks.

From Iran's point of view, the lifting of sanctions allows it to purchase advanced Russian weapons systems, providing Iran far greater military capabilities in the region. Doing so allows Iran to influence the outcome of regional conflicts to its favor since it has the means, the military upper hand, to do so. As a result, Iran gains clout that could be easily managed through economic ventures, specifically contracts with international corporations since the sanctions are lifted. Iran can also gain clout through military alliances with China and Russia, major global powers that wield political influence on the global stage.

Neither Iran's nor Russia's political calculations have been made in a vacuum; rather, they have been well thought out, carefully planned, and soon will be well executed. To coincide with its strategic vision, Iran could not have benefited from better timing to receive the long-overdue delivery of the S-300 air-defense missile system, which Moscow cancelled in 2007 due to the ongoing sanctions on Iran at that time.[14] One concludes how the ongoing upheavals in the Middle East support this idea of a geopolitical shift. To some, the political vacuum in Yemen following the Houthis' seizure of large swathes of the country, the internal discord over power within the Saudi royal family, the sharp drop in oil prices, and, of course, the ongoing civil war in Syria clearly indicate this inevitable geopolitical shift. To others, the maneuvering of Iran and Russia in steering events on the ground suggests that a major confrontation in the Middle East awaits all players. Who is against who? The picture is still blurry as of today; however, it is becoming evident that the Middle East is headed for a proxy cold war in which global powers back the regional players. Undoubtedly, thus far, Iran is strategically well positioned to take advantage of such a cold war. Whether it is to its favor only or that of its supporters, including Russia and China, remains to be seen.

In my opinion, global powers see the shift in the geopolitical landscape of the Middle East as a major threat to global order. Some, like Russia and China, are willing to embrace it, while others, including the United States and Great Britain, want to resist it. As for Iran, it sees itself at the center of this shift because it considers influencing its trajectory, which will inevitably introduce a new political order, at least regionally. Will China and Russia support this theory? My guess is yes because it benefits them politically. That is because Chinese and Russian elites both see the United States as the biggest source of global strategic risk. The average American finds it hard to fathom the latter assertion; however, for one who is well versed in global affairs and geopolitics, who employs in-depth strategic thinking, who is familiar with cultural boundaries under which local politics operate, and who fully comprehends how the ambiguous US foreign policy is indeed contributing to this outcome, the outlook is much different.

Make no mistake: the alliance of Russia and China alone will change the rules of the global political game that has been played for more than half a century. Now imagine adding Iran to the mix, at least in the Middle East. One ends up with an ideological and political showdown between the East and the West. Russia and China can justify their involvement in this geopolitical shift as a response to the United States undermining their strategic interests: "The US' efforts to encroach on China and Russia's strategic room has rendered an interdependence between Beijing and Moscow over some core interest issues."[15]

I see that Russia and China are strengthening their bilateral ties to achieve global strategic stability. That goal is evidenced by both Xi Jinping of China and Vladimir Putin of Russia visiting Iran in a show of support for Tehran while sending a message to the United States, challenging its hegemony in what it considers its own turf—the Middle East. Among the three nations—Iran, Russia, and China—one finds a political flavor for everyone. As I argued in my previous writings, though it may seem odd that bilateral ties between these three countries continue to grow and converge on the goal of undermining the United States' global leadership, each one has its own agenda and goals. At least for China and Russia, it is clear why their relations have thawed lately.

However, the alliance between Iran and its nuclear agreement backers, Russia and China, is one of necessity that will eventually manifest itself beyond the political sphere. Mark McNamee, Central and Eastern Europe analyst at Frontier Strategy, argues, "Of far more importance [than energy] is the political support Russia offers [China], regarding foreign policy matters at the United Nations, Group of 20 and other venues. . . . China, naturally, is happy to have a useful ally as it seeks to reform the existing U.S.-led order to attain its geopolitical goals."[16] Granted, Russia, a former superpower, now tries to regain its status on the global stage, and China, the newest superpower, expanding its military presence in the South China Sea. But what is Iran after?

I am certain countries in the Middle East and elsewhere wonder what this newly formed triangle of Iran, China, and Russia is capable of achieving: Rouhani in the Middle East, Jinping in the South China Sea, and Putin in Europe. Time is of the essence for the West to have its strategy in place to deal with the inevitable shift. This shift goes beyond diplomatic rhetoric, meetings, and political optics. It needs to get to the heart of the matter, to understand the ramifications of this shift's meaning for global order as we know it. It is fair to say that, at least with Iran, the West has a grasp, no matter how limited, of how political decisions are made. It is a different ballgame with China and Russia. Much of the Chinese military and political decision-making process occurs beyond the sight of the outside world. As for Russia, I am reminded of what Winston Churchill once said about Russia: "It is a riddle, wrapped in a mystery, inside an enigma." He might well have said the same of China today.

Let us get an idea of China's strategic vision for the Middle East once Iran acquires nuclear weapons. Consider China's first overseas naval base in none other than Djibouti. It certainly raises serious questions and generates concerns in Western capitals as the naval base will play a pivotal role in China's strategic vision. I argue that China wisely pursues the strategic access provided by this naval base. James Holmes, professor of strategy at the Naval War College, writes,

> Last week, Chinese engineers broke ground on what press accounts styled "China's first overseas naval base" in Djibouti. That is a big deal. Djibouti lies in East Africa along the Bab el-Mandeb Strait, the waterway that connects the western Indian Ocean with the Red Sea. It also adjoins the patrol grounds for the Gulf of Aden counterpiracy mission, in which China's navy has taken part since 2009. In short, it occupies strategic real estate. China, however, will not be the lone occupant of the seaport. Djibouti is also home to other foreign logistics hubs: The Japan Maritime Self-Defense Force operates a facility there, for instance, as does the U.S. Navy. But does the Chinese installation mark a shift in Beijing's naval outlook? Is Djibouti indeed an overseas naval base, the first in the "string of pearls" that has occasioned so much commentary over the past decade?[17]

Coupled with the much-anticipated shift in the geopolitical landscape next door in the Middle East, China is well positioned to influence the course of political events to its favor whether the West likes it or not. China's strategic approach does not end here when considering the ongoing rift surrounding the South China Sea dispute. I do not see how China could give up the islands considering the military infrastructure it built there and vast expansion it embarked on. I go further, stating that when it comes to the South China Sea, Beijing is willing to engage militarily if it means defending its rights over the islands. The latest rebuff from China in response to the International Criminal Court ruling reinforces my point.

Putting China's strategic, assertive foreign policy in the context of its strategic vision and logic, I argue that the establishment of its first naval base in Djibouti and the firm stand regarding the islands in the South China Sea suggest that Beijing has already thought of ways to position itself amid sea lanes throughout the Indian Ocean through which raw materials and finished goods transit. Whether other countries, such as India, might object to China's strategic approach remains to be seen. China could enter into an agreement of some sort with India in which China provides military assistance to New Delhi. I am certain the Indian prime minister, Narendra Modi, would be receptive to China's proposal.

Anticipating a geopolitical shift in the Middle East, with Iran playing a far greater role, China is strategically thinking of building its naval infrastructure around key strategic ports on the Indian Ocean with easy access to the Middle

East. I believe China's endeavors go beyond routine stops at naval bases for fueling and dropping supplies. Rather, China aims to strengthen its posture, sending a message to the United States and others that it will take the necessary measures on land, by sea, and through the air to defend its strategic interests. Establishing the Chinese naval base in Djibouti means that Beijing understands that its newly found global power affords it the opportunity to build a political foundation and forge ties with the countries it believes play a pivotal role in realizing China's strategic vision. Thereby China kills two birds (expanding its naval presence across key strategic locations while laying the foundation to protect its interests should circumstances warrant) with one stone (providing governments where these naval bases are located some form of protection). China's grand strategy did not happen in a vacuum; rather, credit goes to Alfred Thayer Mahan, the foremost sea-power theorist of the nineteenth century. James Holmes writes,

> China, accordingly, has taken its cue in part from Alfred Thayer Mahan, the preeminent sea-power theorist of the late 19th century and the modern U.S. Navy's intellectual founder. Chinese strategists have woven a Mahanian strand into their deliberations, fashioning a strategy best described as an amalgam of Western and Chinese concepts about marine endeavors. Mahan's approach, in which commerce is king, defines sea power in terms sure to appeal to contemporary China. For Mahan, a people's propensity to trade is the chief determinant of its maritime fortunes. Commercial access to important theaters constitutes the uppermost purpose of sea power. Political and military access are mere enablers.[18]

China expects the geopolitical shift in the Middle East to shake up the world order among global powers, redefine priorities among key regional players, and rearrange the seats at the global table among all parties involved. When Iran declares itself a nuclear power, I believe that China will aggressively pursue its geostrategic interests, whatever those may be.

Why is the focus mainly on China but not as much on, let us say, India, Russia, or Germany? The answer lies in my earlier assertion, in which I argue that it is likely that China, rather than Russia or India, will enter into a strategic partnership with Iran that will permit that nation to establish a naval base on either the Arabian or the Caspian sea off the Iranian coast. So, it makes sense to focus on China since the latter will benefit economically, politically, and strategically from the inevitable geopolitical shift in the Middle East.

And how convenient it will be for Beijing to increase its maritime influence given that its base in Djibouti will be fully operational by then. I am interested to see with which country China might consider establishing its next naval base. Arrangements are already in the works given how Iran operates on multiple fronts when it comes to establishing ties with other countries after the lifting of sanctions. That move greatly benefits China. For instance, Iran entered into an agreement with the Indian government, providing it the rights to develop and operate the Iranian port of Chabahar, near the Pakistani port of Gwadar, along Iran's Arabian Sea coast. This would undoubtedly benefit China's long-term strategic objective. Could this explain China's steady expansion in the South China Sea and other areas? China sees it as a gesture of goodwill to lend a supporting hand to local governments just in case one day it might call on those same governments to allow Beijing to establish naval bases. China's strategy came to fruition during the Association of Southeast Asian Nations' (ASEAN) latest meeting in Vientiane, Laos, when the Obama administration tried to urge the ASEAN to make a reference to the July 12 ruling by the UN-backed Permanent Court of Arbitration, in which US ally Manila won an emphatic legal victory over China. It was a big blow to the Americans as Cambodia objected to this reference. Michael Martina writes, "China scored a diplomatic victory on Monday as Southeast Asian nations dropped a U.S.-backed proposal to mention a landmark international court ruling against Beijing's territorial claims in the South China Sea in a joint statement. A weekend deadlock between Association of Southeast Asian Nations (ASEAN) foreign ministers was broken only when the Philippines withdrew its request to mention the ruling in the face of resolute objections from Cambodia, China's closest ASEAN ally. China publicly thanked Phnom Penh for the support, which threw the regional bloc's meeting in the Laos capital of Vientiane into disarray."[19]

It appears that China's forging stronger ties with other countries while establishing naval bases in strategic locations fits well with its long-term strategy. Simultaneously, it sends a strong message to the West—and the United States, in particular—that the political order that once dominated the Pacific is no longer in effect, and the days of US hegemony in the region are numbered. I believe that the same thing could be said about the upcoming geopolitical shift in the Middle East, which I foresee China playing a pivotal role in, shaping its outcome to its favor. That will force a change in the global order of that vital region. The West should not be surprised, as these dynamics are to be expected when a rising power like China subtly expands its military and economic capabilities. The case for the geopolitical shift in the Middle East will be even more interesting, to say the least.

From Russia's Perspective

What about Russia? What could be said about its strategy? Unfortunately, Russia's objectives differ from those of China. The reason is that Russia's overall endgame is to reintegrate the Baltic states following the fall of communism in the early 1990s. Although it is a great aspiration, achieving this objective is a challenging proposition for Putin as these Baltic states that were once in the Soviet orbit do not want to turn back the clock. Yet, Putin sees the geopolitical shift in the Middle East as a conduit for Russia to exert a more assertive foreign policy, thus legitimizing a challenge of the status quo and, if possible, introducing a new global order that recalls Russia's past strength. Will Putin succeed? The answer depends on who holds the geopolitical wild card when the deck of geopolitical cards in the Middle East is reshuffled.

Evidently, one cannot draw a comparison between Russia and China when it comes to the inevitable geopolitical shift in the Middle East. The reason is that Russia's long-term strategic interests do not lie in the Middle East but in the Visegrad countries (the Czech Republic, Hungary, Poland, and Slovakia) in addition to Lithuania, Ukraine, and Slovenia—countries once under the Soviet flag. Yet, turmoil in the Middle East, a geopolitical shift there, and a reassignment of the global powers' roles provide Russia a historic opportunity to move aggressively both tactically and politically, if it has not already done so. Russia's annexation of Crimea shows its understanding of the opportunity. Given the realities in the Middle East, seen through a geopolitical lens, I argue that Russia's strategic aspirations ought to cause great concern in Western capitals. The Middle East is nothing but a platform on which Russia can launch these aspirations. With the civil war in Syria entering its fifth year, showing no sign of abating, and more recently the nuclear agreement Iran reached with the West, Russia is well positioned to influence the geopolitical outcome in a region known for its shifting loyalties and conflicting passions and ideologies.

Russia's recent activities in the Middle East, especially the movement of its heavy weapons to the Syrian theater, suggests that Moscow aims at something far greater than just lending support to the embattled al-Assad regime. Equally important, Russia's political calculations aim at two specific objectives.

First, Putin realizes the role Russia plays in the Middle East by inserting itself in operations: he forces other players, including the United States, to deal with the Kremlin whether they like it or not. Even US Gulf allies, including Saudi Arabia, work behind the scenes to negotiate with Russia to end the Syrian civil war. Moscow communicates to the West in general and the United States in particular that Russia can no longer be perceived as isolated or irrelevant. The Obama administration finds itself in an awkward political position regarding how to clean up the Syrian quagmire. Despite what has been portrayed publicly, Washington and Moscow are working behind the scenes to sort out the various groups and possibly figure out a political solution to the Syrian civil war that meets both parties' strategic interests.

Second, the massive presence of Russian armament in Syria and its targeting of ISIS and rebel groups push debaters in Western capitals to answer the question, what is the exact role of Russia, especially given ISIS's growing threat? Putin's strategy in Syria, whatever it may be, has certainly lent him credibility as the leading force in the anti-ISIS campaign. I believe Russia cornered the West into accepting the political reality vis-à-vis the Syrian conflict that the West benefits politically and operationally by including Russia in any future debate rather than marginalizing it. Washington and Moscow understand this reality more than any capitals in the world. I wonder how long Russia can manage the Syrian conflict alone. At some point, Russia must decide if it is worth chasing an illusion—a return to a pre-Syrian civil-war era. Maxim Suchkov, an expert at the Russian International Affairs Council and a columnist for *Al-Monitor's Russia Pulse*, writes, "All of those developments, however, must be sustained. Given that there are about 150 groups currently on the ground in the Syrian crisis—and the different amounts of leverage that Moscow, Washington, Riyadh, Doha and Tehran have with their respective proxies—practical implementation of a political transition may be impossible. Nonetheless, Moscow's intent to bring the conflict into the political realm as soon as possible seems real and understandable; carrying it out militarily is a politically costly and demanding enterprise, especially when acting alone."[20]

Rhetoric aside, Moscow understands full well that its engagement in the Middle East, especially after Iran's nuclear agreement and the enthusiastic support of Tehran, translates into Iran's support of Russian forces in the Syrian theater. Make no mistake, Russia's long-term strategy, the presence of a military front line as a bulwark against NATO, goes through the Middle East. The chaos in the region provides Moscow the perfect opportunity. Equally important, Russia presumes that religious tensions between Sunnis and Shiites will continue to dominate the religious landscape in the Middle East for years to come. The severance of diplomatic ties between Tehran and Riyadh supports my argument.

Russia's involvement in global affairs, especially after the US invasion of Iraq and Moscow's realization of the outcome of the United States' disastrous adventure there, has been noticeable. It was just a matter of time before Iran expanded its political influence into Iraq, guiding Baghdad's political decisions to the eventual political and strategic benefit of Tehran. That is exactly what happened when the United States started to withdraw its forces from Iraq in 2007 and completed the process in 2011. The security vacuum created by this withdrawal not only allows Iran to move in forcefully but also permits Moscow to redirect and refocus its strategy given those force-producing events. The civil war in Syria was the perfect opportunity to move its strategy from a theoretical phase into an application mode. Moscow's foreign-policy decision to engage militarily in Syria, an area far from Russia's backyard, was unprecedented in its political rule after the fall of the Soviet Union.

I remain convinced that Russia's involvement in the Syrian theater and the presence of its advanced and sophisticated weaponry is to (a) demonstrate its military might and global reach and (b) send a message to NATO that if the latter considers moving eastward, Russia will deal with the issue in NATO's backyard. For this reason, I disagree with those who suggest that Russia's partial withdrawal of its forces from Syria, in March 2016, indicates defeat. Critics who suggested this theory misread Putin's long-term strategy and forget the most basic lesson in international relations: things are not always what they first seem to be. That said, Russia's withdrawal of some of its forces, as some suggested, was not about saving face; rather, the move was nothing more than a drawdown given that, as of this writing, Russia maintains a military presence in Syria. Intelligence imagery indicates the Russian Sukhoi Su-24M bomber aircraft group is still largely intact.

> Russia wants to ensure that the political outcome, whatever form or shape it may be, at least supports some of its long-term objectives. Equally important, an analysis of the imagery shows Russia is not only still expanding infrastructure and facilities around its Latakia naval base but also has deployed additional assets in the past several days. Yes, Mr. Putin caught us all by surprise with his announced Syrian drawdown. Little wonder: Syria-led coalition forces had some momentum behind them; they were on a roll, gaining ground sometimes even without a fight. But make no mistake: This partial withdrawal does not mean that things are settled militarily. The city of Aleppo remains partially encircled by jihad forces who in turn are encircled by coalition forces headed by the Syrian regime. The objective: cut the jihadists' supply lines.[21]

Make no mistake: Russia prefers a political solution to the Syrian conflict, one that I believe Moscow will continue to strongly push for. Recall that President Vladimir Putin stated from the outset that his intervention in Syria had the limited objective to "create conditions for a political compromise."[22] Alas, current conditions suggest that world powers are far from reaching this outcome, and the conflict will most likely drag on for years to come. This, of course, plays out in the hand of key regional players involved as they hope to secure an outcome to the Syrian conflict that is to their liking. My reference is to Saudi Arabia, Iran, and Turkey.

Speaking of Turkey, I argue that, given the recent failed military coup, Erdogan finds it convenient to come down hard on the opposition and also to limit US influence in the region, especially when it comes to the Syrian civil war, by jailing Turkish generals who work closely with Washington. Turkey is capitalizing on the Syrian civil war for political reasons. For instance, Ankara continues to use the issue of Syrian immigrants to blackmail Europe to secure money and, hopefully, gain European Union membership. Alas, the policy of political intimidation and crackdown on the press Erdogan is embarking on shatters hopes for Turkey not only to join the European Union, which I do not foresee, but also to be perceived as a true secular, democratic Muslim state.

For all the saber rattling, Russia wants a political solution to the Syrian civil war, an outcome that must include the al-Assad regime. Iran is pursuing a similar outcome for different reasons. Turkey and Saudi Arabia desire a different outcome that sees not only the removal of al-Assad from power but also the emergence of Sunnis to counter Iran. I argue that the multidimensionality of the Syrian conflict reflects the complicated geopolitical landscape of the region. Sadly, any long-term solution to the conflict is complicated by myriad issues ranging from the erosion of Middle Eastern borders and violence to a flood of refugees and endless sectarian violence.

In this section, my narrative focus on Russia does not mean that other countries, such as the United States, Turkey, Saudi Arabia, Iran, and India, do not have stakes in the outcome of the geopolitical shift in the Middle East. Rather, this focus aims to highlight, from the global powers' perspectives, how the inevitable shift in the geopolitical landscape of the Middle East will influence new dynamics, impact the global order as we know it, and shape the future of international relations.

To Russia's benefit, the ongoing spat between Iran and Saudi Arabia reflects the wider regional conflict between Sunnis and Shiites. The cutoff of diplomatic ties between Saudi Arabia (Sunni dominant) and Iran (Shia dominant) is not new given how proxy wars across the region expose the issue underlying this religious schism between the two Muslim sects. Given this tumultuous environment, one might think that Russia will want to take revenge for Saudi Arabia supporting the Afghan Mujahedeen against the Soviet invasion in 1989. However, this is unlikely because Russia will want to see how the schism between the two religious rivals will play out. Moscow wants to see if the United States will interfere to support its staunch ally, the Saudis, to tilt the cold war to the Saudis' favor. Assuming for a moment that is the case, Moscow has already positioned its heavy armaments in Syria and will be in a better position to shift its priorities and support Iran should the need arise. Alas, the outcome could not be more dangerous.

What about ISIS in this saga? Will it step back and regroup? Or will it side with the Sunnis against the Shiites? Conventional wisdom suggests that, given these uncertain times, ISIS will be more aggressive than ever before, especially when acquiring advanced weaponry. Maxim Suchkov writes, "Besides, if focus moves to the Sunni-Shia nature of the regional conflict, there's a real risk that ISIS could have a respite from the pain it has recently felt, if not a chance at rebirth as a quasi-state with a violent and populist ideology. With the news that Daesh has obtained surface-to-air missile technology capable of downing civil and military aircraft, fighting the Islamic State becomes even more of a challenge for both Russia and the United States."[23]

I am confident that Russia's political elites wonder to what degree Russia's involvement in Syria has cast doubt in the minds of regional Sunni countries about its neutrality. By engaging in Syria, Russia placed itself in an awkward position, being perceived as pro-Shia. As of this writing, Russia strikes me as unconcerned about what impact this perception may have on its political endeavors in the Middle East. Yet, Moscow is embarking on two initiatives to reverse this trend. First, it engages the Kingdom of Saudi Arabia behind the scenes for a solution to the Syrian crisis. Second, Russia accepted Turkey's apology for downing its fighter jet in 2015. This suggests that Moscow is trying to reverse the trend and improve its image among key regional Sunni countries. I have no doubt that Russia's strategy, as the shift in geopolitical landscape in the Middle East takes shape, is one of rapprochement with Sunni countries and a projection of soft power in the greater Middle East. Whether Russia's strategy will work remains to be seen.

The political forecast for the next few years suggests further instability in countries like Yemen, Libya, Iran, Iraq, Lebanon, Egypt, Bahrain, Saudi Arabia, and Turkey, among others. Internal tensions exist among those nations, tensions between Sunnis and Shiites and political tensions between Egypt and Turkey. Speaking of Turkey, I strongly believe the emerging tensions will be fueled by its president's increasingly autocratic style of governance. These ongoing dynamics render the Middle East, more than ever, a fertile ground for a geopolitical shift that transcends the confined borders of the region to manifest itself elsewhere. The picture is even grimmer when adding Russia to the mix—just imagine the chaos that will follow. But let us not kid ourselves; Russia cares less for the schism between Sunnis and Shiites; it cares less about the Middle East; what it cares the most about is how badly Moscow wants to influence the security architecture of the Middle East after Iran acquires nuclear weapons.

For now, Russia promotes its grand vision of stability in the Middle East, a political solution to the Syrian conflict, deescalation between Iran and Saudi Arabia, and the targeting of ISIS. However, the real motive for Moscow's involvement in conducting operations in the Middle East is to contain and prevent the spread of ISIS and the jihadist ideology to Russia's North Caucasus and Central Asia. Current conditions do not favor this particular trend, to say the least.

Fast-forward to when Iran declares itself a nuclear power. Ties between Moscow and Tehran will eventually be vital for sustaining international equilibrium, at least when it comes to who is influencing events on the ground in the Middle East. Still, Iran's and Russia's interests are on opposite ends of the spectrum. Their cooperation, no matter how limited, is essential. I wonder to what extent their relationship remains one of convenience and under what circumstances that relationship could turn bad. For now, suffice it to say that Russia, Iran, and China, for that matter, are pragmatic allies. I am not being naïve when it comes to international relations, especially when addressing both Russia and Iran. I believe that, should Iran's economic influence strengthens to the point of impacting Russia's energy sector, Moscow will have no choice but to become an adversary, though such an outcome is unlikely.

The bottom line is that Russia's anticipation of the geopolitical shift in the Middle East is not about the region but rather the future of the global order and what shape it may take. It will be interesting to learn how the United States intends to address this challenge given its ambiguous foreign policy toward the region. The subsequent section addresses this theme.

The New Geopolitical Realities Postnuclear Iran

Debates are already in full swing as to what the United States' foreign policy toward the Middle East should be as Iran marches toward acquiring nuclear weapons. Some suggest that America's ambiguous policy reflects its lack of understanding of the new dynamics the region continues to endure, including the civil war in Syria, upheavals in Yemen, and sectarian violence in Iraq, among others. Others argue that the American foreign-policy establishment needs a top-to-bottom review that considers adding people from diverse backgrounds who can provide a much better cultural and religious understanding of issues within each country's borders. Since the 9/11 attacks, US foreign policy has been conducted through the barrels of M16s rather than through a clear, objective diplomacy. Said differently, some US policy makers prefer to conduct a militarized foreign policy rather than a policy based on strategic vision—understanding where the world is headed—and a robust diplomatic effort by the Department of State that has within its ranks diversity of thoughts, culture, language, and an understanding of tribal mentality. After all, a diversity of opinions begets creative solutions.

In this section, I address the political role the United States played, and continues to play, in the Middle East. Further, I emphasize what policy, if any, the United States may embark on in the aftermath of Iran declaring itself a nuclear power.

In most of my writings, I present a historical framework to highlight events that affect the course of a particular policy, dealing with foreign governments, or the context of a political agenda, whatever its objectives. In this section, I focus my narrative on what sort of policy or policies the United States needs to embark on when dealing with Iran. Under no circumstance am I suggesting that the United States operates in a political vacuum; rather, I question whether it fathoms core issues and their potential impact.

Take, for instance, the disastrous invasion of Iraq. Given my knowledge and familiarity of the schism between Sunnis and Shiites and of how the invasion of Iraq would remove two of Iran's enemies (Saddam Hussein and the Taliban), I believe that the foreign-policy elites' misguided conviction that America's march toward Baghdad would be a cake walk, as some have suggested, proves the naïveté of those who advocated this policy. The invasion opened old wounds and revived religious tensions between Sunnis and Shiites that have been dormant for so many decades. The rivalry between Iran and Saudi Arabia will drag on for years to come, giving rise to proxy wars in far corners of the Middle East, and Egypt will not find its way back to democracy, no matter how limited. More than ever, the Middle East is like a maze in which the United States has lost itself for lack of an exit, for lack of a path to political stability and safety—and the United States stands to remain trapped, caught in its ambiguous schemes. That is exactly what has taken place within the American foreign-policy establishment.

While events in the Middle East continue to unfold, its borders are becoming blurrier, at least in Iraq, Syria, Libya, Egypt, Jordan, and Turkey. And if may add another country on the horizon: Oman. It appears that this tiny Gulf state is headed for major political turmoil. The reason is that the sultan, Qaboos bin Said, who overthrew his father, Said bin Taimur, in a coup in 1970, has neither son nor brother nor wife to succeed him. This tiny country plays a pivotal political role as a Muslim country not affiliated with either Sunnis or Shiites. Oman also played an important role in bringing American and Iranian negotiators together in 2009 over the latter's nuclear program. Equally important, Oman facilitated the negotiations between the Houthi rebels (backed by Iran) and Saudi Arabia. Unfortunately, these efforts did not yield the desired results. It is easy to overlook this small country and its impact on political Islam, no matter how minor it may be. Kevin Lees, the founder and editor of Suffragio, a website on global electoral politics, writes,

> Increasingly, the Middle East's borders and institutions are grinding through a disruptive reorganization, Westphalian in scale and Hobbesian in brutality. It's easy to overlook Oman, a country of just around 4.4 million (including nearly two million guest workers) that hooks around the southeastern corner of the Arabian Peninsula, and it's even easier to take its stabilizing role for granted. It's certainly at the top of no one's Middle Eastern agenda, in the wake of Turkey's recent failed coup attempt, Syria's ongoing civil war, Iraq's continuing division and the international effort to halt ISIS. Yemen has increasingly become an anarchic quagmire, a pawn in a regional cold war between Saudi Arabia's Sunni kingdom and Iran's Shiite republic.[24]

I remain confident that Iran already has a strategy in place to deal with the aftermath of Qaboos's departure. Interestingly, the sultan who came to power by overthrowing his father some forty-six years ago could face a similar outcome. One thing is for sure: Oman will be part of the geopolitical shift this book highlights. There will be another form of confrontation between Iran and Saudi Arabia. The reason is that, though Oman distinguishes itself from both Sunni and Shia Islam, it remains a close ally of Iran, and therein lies the danger. From my perspective, I can see how political elites in Tehran are already working on a strategy to influence Oman's next political chapter.

There is also another strategic reason that has to do with both oil and limited cooperation—mainly on terrorism—between Oman and the United States. The absence of a successor is already sending mixed messages to interested parties, mainly Iran, Saudi Arabia, and even the United States. I argue that Oman will become another Yemen, another Syria, and another Iraq when it comes to proxy wars between the Muslim rivals (Iran and Saudi Arabia). None of the parties involved will benefit more than Iran given Oman's proximity to the Arabian Sea, access to the Gulf of Aden, and of course the safe transit of oil through the Strait of Hormuz, the narrow passage linking the Persian Gulf to the wider Arabian Sea. The most concerning aspect of the Strait of Hormuz is that Iran, should it decide to close the strait—which I do not foresee unless Iran is militarily attacked—must coordinate its efforts with Oman."

I argue that these dynamics will shape the future geopolitical landscape in the Middle East, and the United States needs to be proactive, not reactive, to events. The civil war in Syria is one example of where the United States must lead.

What can we expect from US foreign policy after Iran's acquisition of nuclear weapons? The answer is anyone's guess. In the subsequent pages, I describe two main scenarios I predict the United States may pursue as part of its strategy in dealing with Iran.

US Containment Policy

First, when Iran declares itself a nuclear power, the United States will need to put its containment policy to work, assuming there is one already in place. Critics who argue in favor of containment should understand that while the policy worked with the USSR (a far more powerful state than Iran), similar tactics may not work with Iran. In this respect, those critics' ideas may be great, methodical, and precise; however, their approach may be headed in the wrong direction and into great danger. *Newsweek*'s Fareed Zakaria and analysts like James Lindsay and Ray Takeyh suggest that the United States can contain Iran and mitigate the danger posed by its nuclear capabilities. Some argue that deterrence against Mao or Stalin might have worked; given that the political landscape then was very different than what it is today, a containment policy against Iran will be useless. If Iran managed to survive through years of hard economic sanctions, what makes these critics think that it will *not* survive containment? Further, Iran is politically better positioned today than it was a few years ago as the sanctions are now lifted, and access to the world could counterbalance the United States' containment policy. Keep in mind that Russia and China, as I stated earlier, will continue to support Iran since they find the Iran nuclear program the best conduit for them not only to gain access to the Middle East through its wider doors but also to challenge US hegemony there.

The United States needs to have a strategy in place not only to deal with the inevitable shift in the political landscape in the Middle East but also to work with other powers whose stakes in the Middle East are in jeopardy. I question, for instance, how the United States might deal with Iran when it comes to the security architecture of the region. My latter assertion in a reference to the emergence of well-coordinated efforts among Iran, Iraq, Syria, and Hezbollah, for example. Yes, we cannot forget the role Hezbollah plays in the Middle East. All these dynamics will eventually impact future political decisions for all parties involved. One sees that the United States needs to have a strategy in place before it is too late. Payam Mohseni, an inaugural director of the Belfer Center's Iran Project and fellow for Iran Studies at the center, writes,

> In a broader picture, aside from Lebanese domestic politics, it is important to watch the emergence of political and intelligence coordination among the self-proclaimed P4+1, composed of Russia, Iran, Iraq, Syria and Hezbollah, focused on shaping new patterns in regional security. From the perspective of the P4+1, the alliance of the United States with Israel and Saudi Arabia is losing its ability to stabilize the Middle East. The future political, military,

and economic decisions in the region will thus have to involve the emerging Iranian-aligned corridor, a loose confederation of strategic actors encompassing more than 100 million people, from Tehran all the way to Beirut.[25]

Consequently, by embarking on a containment policy, the United States will not achieve much because it will be dealing with more than one player. Whatever approach the United States embarks on, there is a high probability that either Russia or China will hinder, if not cripple and undermine, US efforts. Once those countries firmly establish coordination, hostility toward the United States will ensue as rational thinking and pragmatism go out the window, forging a new era in the history of the Middle East, one marked by chaos, military posturing, and policies that guarantee Iran's upper hand in regional affairs.

I have to respectfully disagree with those who argue that the containment-policy approach to a postnuclear Iran will succeed. If Iran sacrificed a lot over the past thirty years of sanctions, what makes proponents of the containment policy think that Iran will yield to US demands? Further, if Iran paid a steep price to achieve its objectives in the region, the political climate postnuclear Iran would only strengthen Tehran's resolve. I believe that an attempt to contain a nuclear Iran would cause it to embrace a more assertive foreign policy. Make no mistake: the West, including the United States and Israel, will react if and when Iran crosses a security threshold. Yet, we ought to ask the questions that matter the most, for example, What are Iran's geopolitical aspirations? Will those aspirations be limited to the Middle East only, or will they go beyond the region's borders? Will Iran's geopolitical calculations change once it acquires nuclear weapons? These are questions the US foreign-policy establishment has to ask and prepare answers for in the hope of not being lost in the political wilderness with many wolves roaming around, waiting for prey.

Undoubtedly, Iran's support of terrorist activities, a tool within its foreign-policy toolbox, is well documented. And there are those who argue that, once it acquires the bomb, Iran will target Israel and US interests in the region. This is nonsense, as these advocates continue to promote extreme, sensational rhetoric that is divorced from reality. I remain convinced that when Iran acquires the bomb, it will pragmatically assess its interests and the mechanisms by which to achieve them. Iran's acquisition of nuclear weapons provide it a much-needed deterrence. Stated differently, the nuclear capabilities Iran will acquire will serve its interests only to a degree; however, we should not lose sight of how, armed with nuclear capabilities, Iran will eventually shape the security and political parameters in the region more so than at any other time. It behooves Iran to demonstrate flexibility, rational thinking, and restraint. The West needs to consider whether Tehran is willing to risk its security and status, at least in the Middle East, by using nuclear deterrence to achieve its geopolitical aspirations. The challenge for Iran is whether it wants to include its ideology (one based on the principles of the revolution) as part of its grand strategy for the Middle East. Doing so could hinder Tehran's efforts, as other regional countries will be far more concerned about what Iran's expansion in the region could mean for their survival. It is likely that Iran's aspirations will be limited to the region given that the rest of the Muslim world has not bought into its revolutionary principles. James Lindsay and Ray Takeyh write,

> The regime has survived because its rulers have recognized the limits of their power and have thus mixed revolutionary agitation with pragmatic adjustment. Although it has denounced the United States as the Great Satan and called for Israel's obliteration, Iran has avoided direct military confrontation with either state. It has vociferously defended the Palestinians, but it has stood by as the Russians have slaughtered Chechens and the Chinese have suppressed Muslim Uighurs. Ideological purity, it seems, has been less important than seeking diplomatic cover from Russia and commercial activity with China. Despite their Islamist compulsions, the mullahs like power too much to be martyrs.[26]

The world has changed, and US foreign policy has to change with it. The containment policy that worked against the USSR will fail against Iran. It is like a tug-of-war game: the party on one end of the rope tries to pull the other from the opposite end. One side represents Iran's aggressive foreign policy; the other end is the West, including the United States. It will be no small matter to contain a nuclear Iran. Similarly, would Iran allow the West to treat it like a pariah since it successfully defied the West when it comes to its nuclear program? Conventional wisdom suggests that with such a success, the mullahs in Tehran will see no reason not to march forward toward joining the nuclear club. Equally important, Tehran has recently learned a few lessons about American foreign policy, namely that it is heavy on rhetoric and light on substance. Stated differently, the United States issues warning to countries but hardly follows through. America's nonresponse to the Syrian regime's attacks using chemical weapons bears out my point. We have yet to learn which country or countries, if any, would join the United States in this containment policy. The United States cannot do it alone since the current political landscape of the region involves other world powers (China and Russia), who see political and economic opportunities to support Iran.

As far as Iran attacking Israel once Iran acquires the bomb—it is nothing but hype. Iran fully understands the ramifications of such an attack. For this reason, Tehran would think twice about its interests before embarking on such a policy in the region. Former president Carter's national security advisor Zbigniew Brzezinski argues that, despite the possibility of danger that Iran exhibits, nothing in the regime's history suggests that it is suicidal.

But we should not be naïve in thinking that if Iran sees an opportunity it will not take it. For instance, the Iranian regime recalls when the Bush administration issued a warning to North Korea about its nuclear program, which turned out to be nothing but a bluff, and when the Obama administration issued a warning for the Syrian president al-Assad that there would be consequences if he launched chemical attacks against civilians. When he did, the United States was politically paralyzed, unable to do anything about it. My point is that this sort of loud rhetoric goes against the spirit of containment if the policy cannot be enforced. If, for instance, I were to advise the Iranian government, all I would have to say is that the United States is paralyzed politically following years of hollow threats, weak leadership, and a disastrous foreign policy. More than ever, Iran and other countries are convinced that all they hear from Washington are empty threats. In their opinion, the United States lost its credibility as its leadership diminished. When an American president gives foreign-policy statements to Iran and other countries, those countries hear only meaningless bravado, absurd prating about global leadership—global leadership being a rare political commodity these days. America lacks resolve, and its greatness in standing for freedom and democracy amounts to nothing but a consumption of ink.

For these reasons, the containment policy the United States might consider for Iran must go through serious debates, setting the objectives and not wavering when things go south. Once again, calls from the White House about what the United States can do to contain Iran would fall on deaf ears. I am certain the mullahs would not take American threats seriously. The danger in this proposition lies in whether the United States alone can carry the burden that comes with the containment policy. Will other US allies in the region back the United States' efforts? Or will they suggest, as they always do, that the United States fight on their behalf against Iranian encroachments? As of this writing, Gulf states and other regional countries (mainly Sunnis) still see the United States as the sole guarantor of their security. But for how long? And at what cost? I am certain the United States would honor its security obligations toward it allies; however, who is to say that one day a Sunni Arab country like Egypt, Saudi Arabia, Morocco, or Jordan could not embark on an indigenous nuclear program of its own? The possibility is there, though reality suggests otherwise.

Let us assume for a moment that some of these Sunni countries, such as Turkey, Jordan, Egypt, and Saudi Arabia, decide to develop nuclear capabilities of their own. Doing so would truly set the region on a collision course not between countries but between terrorist groups and the governments of the nations wherein they reside. One thing is certain: if rogue actors get their hands on a nuclear device, they are not going to think twice about whether to use it. It is a forgone conclusion, to say the least. Should this scenario become a reality, the geopolitical shift in the Middle East would become clearer: supported by a nuclear deterrent, Iran would use its newfound influence to convince other regional countries with majority Shia populations (Kuwait, Bahrain, and Qatar) to cut their ties with the United States and shut down US military bases in their territories. The other scenario is one where regimes like the Saudi monarchy, Egypt, and Jordan collapse from within, either through a coup (as in Egypt) or assassinations. Rogue elements within those states will gain access to facilities where the nuclear devices are stored and move the devices to unknown locations for use at a later date.

As stated in this chapter's introduction, I am trying to present a reasonable scenario regarding what this geopolitical shift in the Middle East might look like in the next thirty years. I remain optimistic that the scenarios listed above will not occur, assuming American intelligence agencies are not chasing their tails. However, if and when these scenarios become reality, the United States will have some difficult decisions to make, including whether to (a) leave the Middle East to its own perils or (b) convince other countries to give full-fledged support for the containment policy. The latter proposition would prove challenging indeed.

The bottom line is that there is an element of risk involved should the United States embark on a containment policy in the aftermath of Iran acquiring nuclear weapons. Knowing that Russia and China support it, Iran might be tempted to see how far it could push the United States. This possibility provides a context in which the United States could break away from its verbal threats and move to real action. Would the containment policy against Iran fail? I say so given the past history of the United States managing similar crises. Bret Stephens writes, "In one sense, this analysis is right: should Iran acquire nuclear weapons, the U.S. will have little choice but to attempt to manage the consequences and contain the fallout. Yet containment would be a strategy resting on the rubble of a decade's worth of failed diplomacy. That unsturdy foundation alone—a compound of indecision, cravenness, and squandered credibility—is one reason why the policy would be likely to fail."[27] With this in mind, I still believe that the other option for dealing with a nuclear Iran is to live with it. The United States said the same thing about Pakistan; yet, it accepted a nuclear Pakistan in the end. The question is whether the current and subsequent US administrations are working on a strategy that considers all aspects to render the containment policy a success. Only time will tell.

I find it unrealistic for Washington to expect full cooperation from Iran as it contains it. Iran challenged the West and successfully overcame the sanctions imposed on it for over thirty years. What makes US policy makers confident that a containment policy will achieve its objectives? Alas, Iran—and the Middle East, for that matter—never ceases to surprise us.

Whatever the outcome may be, given the challenges ahead and the inherent dangers should this policy fail to achieve its objectives, the United States should be prepared to live with a nuclear Iran, as it does with Pakistan and India. The only difference is that a nuclear Iran would shadow the security architecture of a

Middle East that never ceases to surprise us, forcing world powers to change their strategy on short notice.

Living with a Nuclear Iran

The other alternative the US foreign-policy establishment has at its disposal is to learn to live with an armed nuclear Iran as it does with Pakistan, India, and North Korea. I do not intend to downplay this dangerous alternative; rather, I am being pragmatic in questioning what else the United States can do against Iran, which survived harsh sanctions for over thirty years. Despite the restrictions imposed upon it, it was able to influence the outcome. Furthermore, regardless of how much of a pariah it continues to be, Iran continues to march forward toward expanding its sphere of influence in a region known for its shifting loyalties and allegiances.

It is worth devoting a few lines to what lies on the horizon for US strategic interests when Iran declares itself a nuclear power. Global affairs and international security analysts are already asking whether the United States can prevent Iran from acquiring the bomb, and if so, how? The answer is not as simple as one may think. Yet, it is clear enough that, at least from Washington hawks' perspective, the only option is a military strike on Iran. But let us take a step back and really think hard about the consequences of a US military strike on Iran and what that could mean for the security of US interests in the region; those of its allies; stability at large; sectarian violence (already expanding); and how other major powers (mainly Russia and China) could benefit from this chaos. I understand the argument that military strikes on Iran might be the *only* best option given that neither sanctions nor containment may prevent Iran from acquiring nuclear weapons. However, the consequences of strikes far outweigh those, let us say, of the United States building a coalition that includes regional countries, in addition to bringing Russia and China on board to support a containment policy. While there are no guarantees this policy would achieve its objectives, at least there is a framework that could be adjusted to consider the interests of all parties involved.

The other aspect that prevents Iran from acquiring the bomb is something the United States is very familiar with: US support for a regime change by Iranian citizens. However, regime change has already been tested, and it failed. Recall the emergence of the green movement back in 2009, which was quickly put down and fizzled out. In my opinion, a regime change led by civilians will have no impact unless elements of the Iranian military establishment back it, and that is very unlikely given that most, if not all, key government sectors, including the military, are under the control of the theocratic establishment. The danger I see in this scenario is that those who advocate a regime change, assuming some foreign entities fully support it, might have underestimated how the Iranian regime, armed with nuclear weapons, might react to that challenge.

The question is whether the mullahs in Tehran—who will brag about their nuclear capabilities, no matter how limited—will also use those weapons to advance their interests beyond the region's borders. A lesson I learned about international relations: context matters. That the term *nuclear capabilities* is used should be enough to rattle Iran's neighbors, forcing them to their drawing boards to seek strategies to mitigate the inevitable geopolitical shift. In this scenario, there is a high probability that the regime might threaten to deploy some of those assets to quell any domestic challenges. Regimes in the Middle East resorted to similar tactics with the use of chemical weapons in Syria and Iraq. I would not be surprised to see how hard the religious establishment comes down on its opponents and any opposition party. Not doing so amounts to the death of the Islamic revolution, something the current regime is willing to defend at any cost. With access to nuclear weapons, the mullahs in Tehran, if and when they realize the end is near, might be compelled to use them regardless of the consequences.

Let history be our guide. Those who advocate a military approach to prevent Iran from acquiring nuclear weapons understand that embarking on such endeavors does not guarantee the desired outcome, nor does defeat reflect the application of a good policy. Recall the 1982 incident when the unpopular and repressive regime of Leopoldo Galtieri in Argentina invaded the Falklands. The outcome certainly was not what Galtieri expected as the war between Britain and Argentina reached a dangerous level. History also reminds us of a similar outcome in the Balkans when Slobodan Milosevic conducted ethnic cleansing, thinking that such actions would provide political capital for him to rule with an iron fist. Alas, the outcome was nothing like what he expected, as he did not last in office after losing the battle when the United States intervened. Those are lessons Washington hawks could learn from when advocating for a policy, whatever that policy may be.

While on this topic, I would like to devote a few words to what politicians should refrain from addressing when it comes to military strikes. Military strikes on Iran's nuclear facilities will have unforeseen, unimaginable, and undesirable outcomes. It should come as no surprise since most military conflicts end up having disastrous outcomes. Recall that the Bush administration argued that the invasion of Iraq would be swift and short-lived. It turned out to be a disastrous political venture, an ill-conceived military operation, an intelligence failure, and a human catastrophe for the Iraqis, an outcome that changed the political landscape of the Middle East for the worse. Yet, US politicians back then who knew that embarking on this military undertaking could have an unwanted outcome preferred, out of political convenience and cowardice, to remain silent.

History is providing the United States another opportunity. Let us hope the United States does not make the same mistake. One thing is sure, however: Iran is no Iraq. With its military strength, intelligence apparatus, paramilitary outreach, and support of militias, Iran is in a position to create havoc and galvanize Iranians of all stripes against the United States. Critics argue that the US military is far superior and that Iran stands no chance of winning the battle. While I agree with the latter assertion, I focus my argument on the long-term impact US military strikes would have. Should this scenario become a reality, the United States would find itself at the center of an already volatile region: the targeting of its interests in the region, furthering instability in the entire region, and the turning against the United States of Islamist militants empowered by different agendas and ideologies.[28]

And yet, with all the different arguments presented and all the dangers and challenges involved, there are those who suggest that containment is the best option the United States has in dealing with the inevitability of a nuclear Iran. I remain skeptical, however, that such a policy will succeed given that Russia and China would undoubtedly undermine US efforts. Yet, the policy itself cannot succeed without the support of these global powers. And I see no reason either Russia, China, or both would want to support the United States at the expense of their own interests and political influence in the greater Middle East.

Presented with the hard facts that thirty years of sanctions on Iran have not changed Tehran's behavior, I do not see how the United States could realistically manage a containment policy that would require other powerful players to make it effective. Even if the United States decides to embark on this policy, does it possess the credibility it needs following decades of failed policies and diplomatic initiatives? American foreign policy in the last decade or so has been marred by confusion, ambiguity, indecisiveness, and shifting positions, depending on which way the political wind is blowing. Those reasons should be good enough to signal the failure of the containment policy.

All things being equal, the United States might just have to learn to live with a nuclear Iran as it does with Pakistan and India. The only difference is that the United States would have to devote more resources to monitor Iran's activities. Whether other countries would support US efforts remains to be seen. As a result of these new dynamics, I wonder whether this approach might pave the way for the United States and Iran to cooperate over issues of interest to both countries, including the elimination of ISIS or creating political stability in Syria and Yemen, for instance. The possibility is there, though not without challenges from US Gulf allies who might see the rapprochement as a threat to their survival. Tensions between the United States and Saudi Arabia are ongoing, as the latter still vehemently opposes Iran's nuclear agreement with the West.

My sense is that, over the years, the Middle East will get used to living with a nuclear Iran. The setting will not differ from when Southeast Asia learned to live with nuclear archenemies Pakistan and India. Yet, both countries are fully aware of the potential danger should one attack the other with nuclear weapons. The result would be mutual destruction for both nations. The scenario in the Middle East obviously differs given how volatile the region is as it continues to experience unparalleled levels of upheaval. The other reason the case differs in the Middle East is Israel's security concerns. However, I am convinced that Iran will not cross that threshold; doing so would amount to suicide, because I believe Israel—and the United States, for that matter—would attack Iran.

Yet, getting used to living with a nuclear Iran requires adherence to a set of protocols that manage and govern that existence. For instance, would it be protocol for Iran to disclose the number of its nuclear warheads? Would there be a treaty of some sort between Iran and its allies for the sake of verifying nuclear materials? How often would scheduled inspections be conducted to adhere to safety measures? What sort of sanctions might the allies agree on in case Iran violates its nuclear agreement? While these measures make sense, applying them requires full cooperation and vast resources. Under no circumstances do I suggest that living with a nuclear Iran at the heart of the Middle East would be easy. Neighboring countries would most likely live in fear and constant concern. Other countries in the region might court Iran, seeking its support in the nuclear realm. Israel would vigilantly stand watch, as its security depends on what goes on around it.

All these dynamics suggest that the Middle East is moving toward a new security environment, one marked by hostilities, distrust, and constant undermining. What role might the allies play in this new security environment? The answer is anyone's guess. Certainly, regional countries' roles would be weak, noncommittal, and most likely accommodate the shrewd, willful regime next door. All these possibilities hinge on whether the Washington establishment has the political wherewithal to traverse the long road ahead, managing a policy that requires political capital, unlimited resources, and the American people's support. Also, should there be no progress in working out the details of either containment or living with a nuclear Iran, the Washington establishment might do what it does best: retreat as things get difficult.

Living with a nuclear Iran is a risky and dangerous proposition, as no country can predict what Iran might do and how it might behave in moments of both strength and weakness. Using history as my guide, I argue that the Syrian regime launched chemical attacks against its people in a time of weakness when no one predicted it. Would Iran behave similarly against another nation? The possibility is there, but I doubt it. Engaging with nuclear weapons is no small matter. Iran would be basically signing its own execution papers as the response would be strong, swift, and *very* destructive. For all the saber rattling, Iranian political elites ought to have enough sense not to cross the red line, whatever that is.

In either case, containment or living with a nuclear Iran, Tehran's behavior would be closely watched, but for how long? After all, there are those who argue that whatever Iran's behavior may be, it is merely a reflection and function of its past experiences. But how can they say that when Iran never had nuclear weapons before? Iran is not losing sleep over what US policy might be when the former acquires nuclear weapons. I am convinced that Iranian political elites are conversing among themselves, saying: If we managed to endure the hardships of sanctions for over thirty years, what makes the Americans and their allies think they would enforce this policy or that policy on us as we join the nuclear club?

I wonder the same, and the only answer I have for now is that only time will tell. The one thing the Washington establishment needs to be very careful about is not to promote Sunni versus Shiite, moderate versus radical, or Arab versus Persian as Iran strengthens its ties with Russia and China. If the United States thinks it could forge a Cold War–like style template on the region to manage a nuclear Iran, it is utterly mistaken since the new geopolitical landscape is far more complex and multidimensional. The setting transcends Washington's ability to control and influence it.

In the next chapter, I provide my perspective on what the United States needs to do to manage the inevitable shift in the geopolitical landscape of the Middle East postnuclear Iran. The approach requires steadiness, careful strategy, and the formulation of policies that fully consider the complexities, and balance the challenges, of Iran while preserving US strategic interests.

7 Author's Reflections

I AM UNDER NO illusion that the geopolitical landscape of the Middle East will shift in response to Iran's nuclear declaration in precisely the way that I have predicted. Yet, my objective in this book is to present a clear picture, an honest assessment, and a realistic approach to the Middle East's geopolitical framework as Iran continues to evaluate the region's political, social, and economic changes. In a region known for its shifting loyalties and contradictions, a nuclear-armed Iran represents a setback for the United States and its regional allies, particularly given Iran's position at the heart of the story. The anxiety among countries in the Middle East stem from not only fear of ISIS and Iran's growing influence, but also from Russia and China's rising presence in the region and how that might impact the balance of power in the greater Middle East. Add to that the massive drop in oil prices, security and political instability in Egypt, the ongoing civil war in Syria, and enduring upheavals in Libya and Yemen. It should come as no surprise that these dynamics are colliding head-on with US interests in the region.

With Iran marching forward toward acquiring nuclear weapons, Washington will find it more challenging to influence the political outcome in its favor. Further, I argue that the upheavals in the region, mainly in Syria, Yemen, Iraq, and Libya, further complicate the United States' ability to have a clear idea of the possible outcome of such turmoil. The Washington establishment should contemplate unconventional policies. It is time for the US foreign-policy establishment to think outside the box and expand its political horizons, especially in the Middle East. Iran's acquisition of a nuclear weapon will shift the balance of power regionally, if not globally.

Make no mistake: China and Russia will not want to miss this opportunity to multiply their footprints in the Middle East as they pursue political and economic ventures. Is it fair to say that the current political turmoil in the region differs markedly in style and substance from that of the Cold War era? The answer is yes because (a) during the Cold War, the enemy, the USSR, was well defined, and (b) the current political climate provides no clear reading of where and how events might evolve as Iran attempts to acquire nuclear weapons.

The United States faces challenges beyond the Middle East. Those challenges extend, for instance, to Asia with ongoing tensions vis-à-vis South China Sea Islands and North Korea's nuclear program endure. Yet, challenges within the greater Middle East persist. Afghanistan represents such a trial where, after fifteen years of military engagement, the United States has yet to show progress.

Pakistan represents another challenge within the greater Middle East, a flash-point for the United States. What makes the case of Pakistan interesting and of concern in this discussion is that, should Saudi Arabia reach out to Islamabad for nuclear assistance, Pakistan's involvement would be a game changer, impacting geopolitical dynamics in the Middle East. The United States would have to have a policy in place that deals with new scenarios. For instance, Pakistan could extend its nuclear umbrella to the desert kingdom. Would India, then, reach out to Iran as a counter strategy?

Given those dynamics, I do not foresee Israel sitting idly by and doing nothing, especially when it finds its security immediately threatened. Would Israel conduct preventive strikes if it becomes evident that Saudi Arabia will move forward with acquiring nuclear weapons from Pakistan? The possibility is there, though it's unlikely. However, should the United States intervene to convince the Saudis *not* to move forward to acquire nuclear weapons, would Israel refrain from conducting strikes? If tensions between Riyadh and Washington persist, I do not see how the Saudis will follow US policy makers' advice. Those dynamics will undoubtedly affect US strategic interests in the Middle East. Needless to say, an ill-defined strategy and an ambiguous policy to address this geopolitical shift could threaten American security and ultimately drown its flailing global leadership.

The issue for the United States does not rest in pursuing a particular policy but rather includes its understanding of how to carry out that policy, whatever it consists of. For this reason, I foresee two plausible policy tracks the United States could consider when and if it decides to deal head-on with this inevitable geopolitical shift.

The first track consists of the US foreign-policy establishment's understanding that, whatever policy the United States decides to pursue, establishing co-operation with Iran will be necessary to its success. Yet, if Iran and the United States decide to cooperate in order to avoid a nuclear disaster in the Middle East, they should not expect everything to go smoothly from the beginning. For instance, Iran fully understands that it has the means and political will to influence whether its agreement with the West over the nuclear issue has prevented a war or merely postponed it. Iran's political elites reason that, if within ten or fifteen years the Islamic Republic will be free to use far more advanced and efficient centrifuges while enriching uranium at a faster pace, it makes sense for Tehran to provide limited cooperation to the United States. Thus, Iran would demonstrate to the rest of the world that it is holding to its end of the bargain.

As this cooperation evolves, I am confident the mullahs will keep a watch on domestic opposition groups, preventing them from rising up against the central government. After all, the possibility of an uprising is there since the lives of average Iranians have not improved since the lifting of sanctions. Iran's hard-liners therefore are not taking any chances that would allow demonstrations to grow and demonstrators to get funding and other support from foreign entities. Yet, the mullahs fully understand that most Iranians, mainly the youth, are eager to join the rest of the world, access global markets, and use technical innovations. I argue that the threat to the Iranian central government is not military strikes against its nuclear facilities; rather, it is the uprising of its youth as the principles of the revolution of 1979 are on life support. The youth want a different life than that of the older generation, whose focus was on studying theology, religious teachings, and the preaching of the Shia principles—to the calculated exclusion of the hard sciences, including mathematics.

Iran could play another card as part of its cooperation: the ISIS card. Playing the ISIS card would be significant given the organization's status as the most reviled terrorist group not only in the Middle East but around the world. Given that ISIS threatens both Iranian and US interests in the region, Tehran wants to capitalize on how ISIS has targeted and beheaded Shiites, thus justifying Iran's argument that ISIS, an extremist Sunni group, threatens Iran and fellow Shiites. Simultaneously, Tehran understands how to play the game, seeing ISIS as a convenient bogeyman. Thus, when Iran, not Saudi Arabia or Turkey, targets ISIS, Muslims in the Middle East perceive Iran as a responsible force in the region. Iran is trying to project to the outside world that it can be a far better partner than Turkey, Jordan, or Saudi Arabia. While the means do not justify the ends, Iran's political calculations *are* to force a political wedge between the United States and Saudi Arabia. Iran sees ISIS as the perfect cause on which Washington and Tehran can cooperate. This tactic could prove useful for Iran's long-term goal: to minimize and limit the United States' role as the geopolitical shift in the Middle East unfolds.

Once again, whether pursuing a containment policy or adjusting to life with a nuclear Iran, the United States will find it challenging to maintain sole control over events in the Middle East in the next few years given how the ongoing turmoil in the region has already shifted the political landscape while the players have multiplied. For all the saber rattling, Washington would be better served by defining its objectives toward either containing or living with a nuclear Iran. More than ever, Washington is urged to develop that strategy now rather than make it up on the fly later.

The second track of this cooperation involves other major powers, mainly China and Russia. While the United States might try to control Iran's nuclear activities, either through inspection or other measures, obtaining the cooperation of those two countries will be a challenging proposition. Under the lens of reality, I do not see how Russia or China will assist the United States given the latter's sanctions imposed on Russia over Crimea and its harsh stance on China over the South China Sea islands. All things being equal, both Russia and China see an opportunity to undermine US efforts, knowing full well that their cooperation in managing Iran's nuclear activities is a forgone conclusion. Washington hawks argue that the United States does not need China—or Russia, for that matter. I have to challenge this assertion, arguing that it was Russia and China that influenced Iran to agree to a nuclear deal with the P5+1. The world has moved on, and Iran has already increased its economic and military ties with both countries. Once again, I do not see how Russia and China would turn back the clock to support the United States at the expense of their own interests unless the latter lifts sanctions on Russia and deescalates its rhetoric over the South China Sea dispute.

But let us take a step back and evaluate why Russia and China will not cooperate with the United States when it comes to containing Iran after it acquires nuclear weapons. The short response lies in the new geopolitical outlook where China and Russia manage to influence Iran's trajectory and hinder US efforts as part of that new environment.

In the case of China, for instance, Washington and Beijing already do not see eye to eye on a host of issues; thus, containing a nuclear Iran would prove to be a divisive issue, and China most likely would limit its cooperation with the United States regarding that concern. The reason is that Beijing will not want to jeopardize its trade with the region, mainly Iran, following the lifting of sanctions. Further, by establishing a naval base in Djibouti, China aims to increase its influence in the region. Iran provides that platform. It is hard to imagine the new security architecture embedded in the geopolitical landscape of the Middle East after Iran acquires the nuclear bomb; yet, the dynamics mentioned herein make that a possibility.

As I argued in my previous writings, it is only a matter of time before China and Iran forge close military ties that could counterbalance those of the United States with Saudi Arabia or of the United States with Bahrain. Iran's move to advance its ballistic missile technology supports my argument. However, I do not think that China worries much about its status as a member of the Nonproliferation Treaty, which requires it neither to provide assistance nor encourage nonnuclear-weapons states like Iran to develop or acquire nuclear weapons. China's influence will not come from this angle; rather, it will derive its stance from *not* cooperating with the United States in containing Iran once the latter joins the nuclear club. China's global influence and prestige rest, to some degree, on how much influence it wields in the geopolitical environment created in the Middle East after Iran gets nuclear technology. Hence, it is to China's political advantage to take a more active role on the global stage, though not yet an assertive one. However, there is no evidence to suggest that China worries too much about what the United States thinks or how it might react to Beijing's flexing its political muscle in a Middle East with Iran as a nuclear power. It all depends on the outcome of the ongoing upheavals.

The United States needs to understand that, despite its rhetoric that Iran is the only source of the challenges emerging from the Middle East (a claim with which I disagree), the main challenge emanates from the outcome of the ongoing regional upheavals. Most of the turmoil is at the center of the greater Middle East

(Iraq, Yemen, Lebanon, and Syria). The Washington establishment must formulate a strategy that ensures that the outcome does not impact US strategic interests. Yes, dealing with a nuclear Iran is no small matter, but the United States has far too many issues to consider that are directly linked to its strategic interests and that transcend Iran's rhetoric.

American foreign policy will be tested like never before over the next few years. The United States will face an uphill battle to understand the new geopolitical setting. It will also witness shifting loyalties and alliances as political, social, and economic conditions in the Middle East change by the hour, where ambiguity, chaos, and violence are the orders of the day. Even those who argue that the ongoing conflicts will eventually stop and that the Middle East may regain some normalcy have not looked far enough to realize that that is not the case. Regional states in the Middle East are falling one by one as though in a domino effect. Nonactor states are growing in strength and global outreach, and the region's complexity makes it harder for the Washington establishment to grasp its depth. How could it when, even from within, the US foreign-policy establishment lacks diversity of thought and personnel. Those with different backgrounds, those who are culturally and linguistically capable, simply are not part of the debate. Not everyone has to conform to the rigid thinking of Washington. Furthermore, a *workable* foreign policy makes flexibility a necessity. My point is that before American foreign-policy makers react to the new geopolitical shift in the Middle East, they ought to consider that the strength of governments is no longer limited to how much they can control but rather to the *key entities* they seek to control.[1]

To frame this within the context of Iran's regional aspirations, I argue that the United States needs to decide how it intends to pursue its foreign policy postnuclear Iran. Will US policy be based on Iran's actions (i.e., violations of the nuclear agreement)? Or will the United States pursue a policy marked by cooperation if Iran changes its policies and practices that the United States finds perilous? Knowing that its two biggest enemies, Saddam Hussein of Iraq and the Taliban of Afghanistan, are eliminated now enables Iran to showcase its strength and appeal to other Shiite minorities scattered around the Middle East. That is exactly what is making the Kingdom of Saudi Arabia nervous, since most of its population in the eastern province (*ash-Sharqiyyah*), mainly in al-Qatif region is Shiite. Could this explain the proxy wars both Saudi Arabia and Iran continue to engage in? I believe so. Add to that the possibility of a rapprochement between Iran and the United States. The Saudis' reaction is anybody's guess.

I am compelled, however, to challenge an assertion from Mohammad Shafiq Hamdam, a writer, political analyst, and Nobel Peace Prize nominee, who claims that the "strategic intentions of Iran include Shiite Islam domination of the region, and the isolation of Israel and Saudi Arabia."[2] My counterclaims are that (a) Iran would not be in a position to isolate Israel and Saudi Arabia given the strategic alliance and signed security treaty each has with the United States, and (b) the domination of Shiite Islam in the region is unlikely given that Sunnis outnumber Shiites. Furthermore, given sectarian violence, I do not see how Sunnis, be they in Turkey and Saudi Arabia or Jordan and Egypt, would allow such domination to happen. Needless to say, Washington hawks made similar claims during Iran's nuclear negotiations with the P5+1. My question then was how could a Shiite population of 170 million around the world control a Sunni population of 1.6 billion? It is nonsense. However, I agree with Hamdam that Iran's aspiration to become a regional power that wields influence and exhibits military strength merits consideration.

This leads me to wonder how US foreign policy toward a nuclear Iran might address a regime that not only functions under a centralized power but also controls key government sectors (the military, security services, and the economy). And do not forget about the lack of freedom of the press, censored information, and the complete absence of freedom of speech. Since American foreign policy is marked by inconsistencies and double standards, I do not know if it would matter much to Washington that Iran *represses* the freedom of its citizens. Recall that the United States turned a blind eye to the military coup in Egypt. Remember that the United States failed to condemn the Bahraini government's use of excessive force against peaceful demonstrators. Iran understands that it has the upper hand, politically speaking. And with the support of Russia and China, Tehran is well positioned to influence the political scene in the new geopolitical landscape to its advantage. How far will it go? That remains to be seen.

The balance of power in the greater Middle East is already tilting, and not to the United States' favor. The promotion of democratic values and principles that the United States holds dear (in words not deeds) will eventually fall on deaf ears as the United States lacks credibility and its leadership suffers major deficits. Yet, I remain hopeful that certain rights (freedom of the press, freedom of expression, and so forth) echo among the Iranian youth, who might rise up against the tyranny of the mullahs and demand the much-needed political change.

Where from here? I find myself asking many questions that I would not have considered a few years ago. For instance, will the geopolitical shift in the Middle East give rise to political Islam? Will the regional order as we know it be a thing of the past given that the rise of Islamists is shaping the political landscape? When you set this narrative before Iran's vision for the Middle East, what you get is the expansion of a Shiite crescent that includes Lebanon, Syria, Iraq, Yemen, Iran, and Bahrain. The United States ought to think in terms of what that expansion means for the development, application, and reception of its foreign policy in the coming years. Mind you, the United States is already suffering from political deficits that include declining support from Arab allies in the region, tensions with Saudi Arabia, an inability to influence the outcome of the civil war in Syria, Iran's ongoing ballistic missile development, a direct challenge from Turkey (a US ally that was supposed to support and further US interests in the region), and an Iraq that continues to be marred by sectarian violence with no end in sight.

These events share common bonds, namely that they (a) impact America's political vision in the region (promotion of democracy, although countries in the region do not believe that anymore), (b) provide economic interests as China and Russia are investing there, and (c) lead to the structure of a new security architecture. So, with the inevitable shift of the geopolitical landscape in the region post-nuclear Iran, Washington will have to make some difficult choices concerning how to apply whatever policy it decides to undertake. The ongoing regional upheavals in addition to the quagmire in Afghanistan and the growing footprints of Russia and China in the Middle East will impact US strategic interests, whether it likes it or not. But I will not go as far as saying that these events will affect the security of the United States. I am certain the United States has taken security measures just in case events in the region challenge its hegemony there.

The Middle East is bound to experience a major political shift once Iran becomes a nuclear power. The changes might not be to the liking of the West, especially the United States. I predict that the region will enter into a new phase of its existence, one marked by even more political turbulence, a new political order, sectarian tensions, and religious rivalry. The region will also witness the rising presence of global powers like Russia and China, whose roles there, throughout history, have been somehow limited in comparison to the United States'.

On the flip side, the region will also experience a new wave of troubles because of the ongoing upheavals. This outcome will certainly reshape both the political outlook and geographical boundaries. A nuclear Iran will also result in new alliances forming along religious, sectarian, and, yes, ideological lines. As unlikely as it might sound, the region will also witness increasing dialogue between the United States, Russia, China, and Iran. The dialogue will consist of the new security architecture set upon a political landscape already shaped by the geopolitical shift. My guess is that the dialogue will include possible negotiations over how to readjust the strategic outlook and interests of all parties involved.

Time is of the essence. US policy makers drawing on leadership from both political parties must put the future of America's strategic interests before party loyalty and resolve to act. If the geopolitical shift in the Middle East is not addressed immediately, accordingly, vigorously, and competently, the consequences will be dire. As the sun rises on the horizon of a troubled region, I wonder whether the West will awaken soon and say "Good morning, Iran."

Notes

1. Introduction

1. David Oualaalou, "Iran's Nuclear Agreement Divides Our Government at Its Core; Allies Won't Care," *Huffington Post*, August 3, 2015, http://www.huffingtonpost.com/david-oualaalou /irans-nuclear-agreement-divides-our-government-at-its-core-allies-wont-care_b_7922028 .html.

2. David Oualaalou, "Geopolitics Is Helping You Top Off the Tank in Your Car," *Waco Tribune-Herald*, January 18, 2015, http://www.wacotrib.com/opinion/columns/board_of _contributors/david-oualaalou-board-of-contributors-geopolitics-is-helping-you-top/article _66756762-3ba1-5d7b-807e-acdb08d91d62.html.

3. Margot Patterson, "The Proxy Wars of Arabia," *America* 212, no. 18 (May 25, 2015): 12, *MasterFILE Premier*, 1–2, EBSCOhost (accessed August 16, 2015), Accession# 102754687.

4. David Oualaalou, "Growing Sunni-Shiite Rift Fuels Power Play, Atomic Pursuit across Middle East," *Waco Tribune-Herald*, May 31, 2015, http://www.wacotrib.com/opinion/columns /board_of_contributors/david-oualaalou-board-of-contributors-growing-sunni-shiite-rift-fuels /article_52a144c3-6bc8-5247-aa87-de3205cef5e3.html.

5. Pew Research Center, "Many Sunnis and Shias Worry about Religious Conflict," *Pew Religious and Public Life*, November 7, 2013, http://www.pewforum.org/2013/11/07/many-sunnis -and-shias-worry-about-religious-conflict/.

6. Lolita C. Baldor, Ken Dilanian, Vivian Salama, and Sameer N. Yacoub, "Iran's Influence in Iraq Solidifies as It Eclipses US in Helping Fight Islamic State Militants," *Associated Press*, January 12, 2015, http://www.foxnews.com/world/2015/01/12/iran-influence-in-iraq-solidifies -as-it-eclipses-us-in-helping-fight-islamic.html.

7. "Background: How Big Is Iran's Military?" *Reuters/Military*, September 28, 2009, http:// www.haaretz.com/news/background-how-big-is-iran-s-military-1.7084.

8. Michael J. Totten, "The Iran Delusion: A Primer for the Perplexed," *World Affairs* 178, no. 2 (Summer 2015): 5–12, *Academic Search Complete*, 2, EBSCOhost (accessed August 16, 2015), Accession# 108420639.

9. Henry M. Paulson, *Dealing with China: An Insider Unmasks the New Economic Super-power* (New York: Hachette, 2015).

10. Beibei Bao, Charles Eichacker, and Max J. Rosenthal, "Is China Pivoting to the Middle East?" *Atlantic*, March 2013, http://www.theatlantic.com/china/archive/2013/03/is-china -pivoting-to-the-middle-east/274444/.

11. Charles L. Glaser, "Time for a U.S.-China Grand Bargain," *Belfer Center for Science and International Affairs*, July 2015, http://belfercenter.ksg.harvard.edu/publication/25586/time _for_a_uschina_grand_bargain.html.

12. David Oualaalou, "Yemen Betrays US Military Decline in a Chaotic Middle East," *Huffington Post*, April 7, 2015, http://www.huffingtonpost.com/david-oualaalou/yemen-betrays-us -military_b_7011856.html.

13. David Oualaalou, "China's Leadership Role on Global Stage Will Continue to Grow," *South China Morning Post*, July 22, 2015, http://www.scmp.com/comment/letters/article /1842787/chinas-leadership-role-global-stage-will-continue-grow.

14. Charles L. Glaser, "A U.S.–China Grand Bargain? The Hard Choice between Military Competition and Accommodation," *International Security*, May 1, 2015, http://www.mitpress journals.org/doi/abs/10.1162/ISEC_a_00199#.VbxUTvmYl7y.

15. "Middle East and China: Remarks to a Conference of the United States Institute of Peace," *Middle East Policy Council*, February 17, 2015, http://mepc.org/articles-commentary /speeches/middle-east-and-china.

16. Christina Lin, "The New Silk Road: China's Energy Strategy in the Greater Middle East," *Washington Institute for Near East Policy*, April 2011, http://www.washingtoninstitute .org/uploads/Documents/pubs/PolicyFocus109.pdf (12).

17. Stephen Eisenhammer, "More Countries Say to Join China-Backed AIIB Investment Bank," *Reuters*, March 28, 2015, http://www.reuters.com/article/2015/03/28/us-asia-aiib -china-idUSKBN0MO00F20150328.

18. David Oualaalou, "US Era of Dominance Is Dwindling as China Takes Over the World Economy," *Huffington Post*, December 15, 2014, http://www.huffingtonpost.com/david -oualaalou/us-era-of-dominance-is-dw_b_6299040.html.

19. "China's Hypersonic Strike Vehicle in 3d Test Flight," *RT Network Global News*, December 4, 2014, http://www.rt.com/news/211575-china-hypersonic-missile-test/.

20. Lidia Kelly, Denis Pinchuk, and Darya Korsunskaya, "India, Pakistan to Join China, Russia in Security Group," *Reuters*, July 10, 2015, http://www.reuters.com/article/2015/07/11 /us-china-russia-idUSKCN0PK20720150711.

21. "Iran Watering Down Russian Trade Deal Terms," *Stratfor Analysis* (August 1, 2014): 31, *Business Source Complete*, EBSCOhost (accessed August 16, 2015), Accession# 97892084.

22. David M. Herszenhorn and Michael R. Gordon, "U.S. Cancels Part of Missile Defense that Russia Opposed," *New York Times*, March 16, 2013, http://www.nytimes.com/2013/03 /17/world/europe/with-eye-on-north-korea-us-cancels-missile-defense-russia-opposed .html?_r=0.

23. Stephen Blank, "Russia's Return to the Middle East," *World & I* 11, no. 11: 314, *Master-FILE Premier*, EBSCOhost (accessed August 16, 2015), Accession# 9610314661.

24. David Oualaalou, "NATO's Invitation of Montenegro Betrays Folly and Lack of Strategic Vision," *Huffington Post*, December 14, 2015, http://www.huffingtonpost.com/david -oualaalou/natos-invitation-of-monte_b_8801886.html.

25. David Oualaalou, "Iran's Nuclear Agreement Divides Our Government at Its Core; Allies Won't Care," *Huffington Post*, August 3, 2015, http://www.huffingtonpost.com/david-oualaalou /irans-nuclear-agreement-divides-our-government-at-its-core-allies-wont-care_b_7922028 .html.

26. David Oualaalou, "Russian Ire, US Strife Could Torpedo Fragile Nuclear Negotiations with Iran," *Waco Tribune-Herald*, February 26, 2015, http://www.wacotrib.com /opinion/columns/board_of_contributors/david-oualaalou-board-of-contributors-russian -ire-us-strife-could/article_4f964b6f-dfob-50b4-bfd3-4217e6c91b04.html.

27. Natasha Bertrand, "One of the Biggest Weaknesses in the Fight against ISIS Is Being Fully Exposed in Turkey," *Economic Times*, August 1, 2015, http://economictimes.indiatimes .com/articleshow/48308010.cms?utm_source=contentofinterest&utm_medium=text&utm _campaign=cppst.

28. Burak Bekdil, "What Turkey Wants in Syria," *Middle East Forum: The Gatestone Institute*, July 31, 2015, http://www.meforum.org/5415/what-turkey-wants-in-syria.

29. David Oualaalou, *The Ambiguous Foreign Policy of the United States toward the Muslim World: More than a Handshake* (Lanham, MD: Rowman & Littlefield, 2016).

30. Frida Ghitis, "Saudi Arabia and Iran Face Off in Yemen," *World Politics Review (Selective Content)* (November 26, 2009): 1, *Academic Search Complete*, EBSCOhost, Accession# 80304065.

31. Mahmud A. Faksh, *The Future of Islam in the Middle East: Fundamentalism in Egypt, Algeria, and Saudi Arabia* (Westport, CT: Praeger, 1997).

32. David Oualaalou, "What Are the Saudis and Hamas Planning Behind Our Back?" *Huffington Post*, August 10, 2015, http://www.huffingtonpost.com/david-oualaalou/what-are-the-saudis-and-h_b_7962308.html.

33. Reback Gedalyah, "Will Egypt Go for Its Own Nuclear Weapon?" *The Middle East*, March 2015, http://www.israelnationalnews.com/News/News.aspx/192092#.VcmEUPmYl7w.

34. Polina Tikhonova, "Russia and Saudi Arabia Outplayed U.S. in the Middle East," *Politics*, August 2015, http://www.valuewalk.com/2015/08/russia-and-saudi-arabia-outplayed-u-s-in-the-middle-east/.

35. David Oualaalou, "Warming Saudi, Russian Ties Unlikely to Endure Geopolitical Tumult in Middle East," *Waco Tribune-Herald*, July 16, 2015, http://www.wacotrib.com/opinion/columns/board_of_contributors/david-oualaalou-board-of-contributors-warming-saudi-russian-ties-unlikely/article_bd7126ee-cfe5-5551-9f64-8b22cb349a7e.html.

36. Sadiq Alkoriji, "Hugh Kennedy, When Baghdad Ruled the Muslim World: The Rise and Fall of Islam's Greatest Dynasty," *Library Journal* (2005): 101, *Literature Resource Center*, EBSCO*host* (accessed August 15, 2015), Accession# edsgcl.132418583.

37. "Justice and Development Party (AKP)," *Global Security*, August 2013, http://www.globalsecurity.org/military/world/europe/tu-political-party-akp.htm.

38. John Feffer, "The Breakup," *Foreign Policy in Focus*, May 18, 2010, http://fpif.org/the_breakup/.

39. David Oualaalou, "Iran's Nuclear Agreement Divides Our Government at Its Core; Allies Won't Care," *Huffington Post*, August 3, 2015, http://www.huffingtonpost.com/david-oualaalou/irans-nuclear-agreement-divides-our-government-at-its-core-allies-wont-care_b_7922028.html.

40. Jordan Chandler Hirsch, "How America Bamboozled Itself about Iran," *Commentary* 139, no. 4 (March 1, 2015): 16–22, https://www.commentarymagazine.com/articles/how-america-bamboozled-itself-about-iran-1/.

2. History of the Persian Empire

1. "A Brief History of Persian Empire," December 20, 2015, http://www.parstimes.com/library/brief_history_of_persian_empire.html.

2. Ken Nelson, "Ancient Mesopotamia: Persian Empire," *Ducksters*, December 20, 2015, http://www.ducksters.com/history/mesopotamia/persian_empire.php.

3. Darris McNeely, "Alexander the Great: The Man Who Would Be God," United Church of God, January 30, 2005, http://www.ucg.org/the-good-news/alexander-the-great-the-man-who-would-be-god.

4. Ervand Abrahamian, *A History of the Modern Iran* (Cambridge, UK: Cambridge University Press, 2008), 35.

5. Ervand Abrahamian, *A History of the Modern Iran* (Cambridge, UK: Cambridge University Press, 2008).

6. Ibid., 97.

7. "History of Iran: Reza Shah Pahlavi, the Great," *Iran Chamber Society*, April 5, 2016, http://www.iranchamber.com/history/reza_shah/reza_shah.php.

8. Ervand Abrahamian, *A History of the Modern Iran* (Cambridge, UK: Cambridge University Press, 2008), 98.

9. Abbassi Zahra, "Anglo Iran 1919 Agreement," *Iran Review*, July 23, 2012, http://www.iranreview.org/content/Documents/Anglo-Iran-1919-Agreement.htm.

10. "March 25, 1946: Soviets Announce Withdrawal from Iran," *History*, 2009, accessed April 6, 2016, http://www.history.com/this-day-in-history/soviets-announce-withdrawal-from-iran.

11. Ervand Abrahamian, *A History of the Modern Iran* (Cambridge, UK: Cambridge University Press, 2008).

12. Takeyh Ray, "What Really Happened in Iran: The CIA, the Ouster of Mosadeqq, and the Restoration of the Shah," *Council on Foreign Relations*, July/August 2014, http://www.cfr.org/iran/really-happened-iran/p33125.

13. Ervand Abrahamian, *A History of the Modern Iran* (Cambridge, UK: Cambridge University Press, 2008).

14. "1953 US Coup in Iran and the Roots of Mideast Terror," *Rense*, August 23, 2003, http://rense.com/general40/roots.htm.

15. Ervand Abrahamian, *A History of the Modern Iran* (Cambridge, UK: Cambridge University Press, 2008).

3. Emergence of Modern-Day Iran

1. Michael Axworthy, *A History of Iran: Empire of the Mind* (New York: Basic Books, 2010).

2. Ibid.

3. Hashemzadeh Kianoosh, Craig Belanger, and Alex K. Rich, "Iran," *Salem Press Encyclopedia: Research Starters*, January 2014, EBSCOhost (accessed February 3, 2016), Accession# 88391100.

4. Ibid.

5. Ali Alfoneh, "The Basij Resistance Force," United States Institute of Peace, 2010, http://iranprimer.usip.org/resource/basij-resistance-force.

6. Hooman Majd, *The Ayatollah Begs to Differ* (New York: Random House, 2008), 115.

7. Ali Alfoneh, "The Basij Resistance Force," United States Institute of Peace, 2010, http://iranprimer.usip.org/resource/basij-resistance-force.

8. The World FactBook. *Central Intelligence Agency*. Retrieved from https://www.cia.gov/library/publications/resources/the-world-factbook/geos/ir.html.

9. "Persepolis—the Golden City of Achaemenid Empire," *Tripfreakz*, May 18, 2015, http://tripfreakz.com/offthebeatenpath/persepolis-achaemenid-empire-ancient-persian-iran.

10. "The World's Largest Cities and Urban Areas in 2006," City Mayors Statistics, September 25, 2010, http://www.citymayors.com/statistics/urban_2006_1.html.

11. "The Sunni-Shi'a Divide," Council on Foreign Relations, 2014, http://www.cfr.org/peace-conflict-and-human-rights/sunni-shia-divide/p33176#!/p33176.

12. Herb Shapiro, "Light and Shadows: The Story of Iranian Jews," *Library Journal*, 2013, 79, *Literature Resource Center*, EBSCOhost (accessed February 15, 2016). edsgcl.324980003.

13. Jamsheed K. Choksy, "Non-Muslim Religious Minorities in Contemporary Iran," *Iran and the Caucasus* 16, no. 3 (October 2012): 271–99. *Academic Search Complete*, EBSCOhost (accessed February 15, 2016), Accession# 83412815.

14. "The Sunni-Shi'a Divide," Council on Foreign Relations, 2014, http://www.cfr.org/peace-conflict-and-human-rights/sunni-shia-divide/p33176#!/p33176.

15. "Guardianship of the Islamic Jurists," *Abbreviations*, accessed March 2016, http://www.abbreviations.com/GUARDIANSHIP%20OF%20THE%20ISLAMIC%20JURISTS.

16. David Oualaalou, "Growing Sunni-Shiite Rift Fuels Power Plays, Atomic Pursuits across Middle East," *Waco Tribune-Herald*, May 31, 2015, http://www.wacotrib.com/opinion/columns/board_of_contributors/david-oualaalou-board-of-contributors-growing-sunni-shiite-rift-fuels/article_52a144c3-6bc8-5247-aa87-de3205cef5e3.html.

17. "Guardian Council/Council of Guardians," Global Security Organization, September 7, 2011, http://www.globalsecurity.org/military/world/iran/guardian.htm.

18. Barbara Ann Flanagan-Rieffer, *Evolving Iran: An Introduction to Politics and Problems in the Islamic Republic* (Washington, DC: Georgetown University Press, 2013), 47.

19. James Buchan, *Days of God* (New York: Simon & Schuster, 2013), 95.

20. Shalaleh Zabardast, "Flourishing of Occidentalism in Iran after Cultural Revolution," *Journal of Gazi Academic View* 9, no. 17 (December 2015): 215–28, *Academic Search Complete*, EBSCO*host* (accessed February 3, 2016), Accession# 11891854.

21. "Iran 2015 Country Review," *Iran Country Review*, July 2015, 1–479, *Business Source Complete*, EBSCO*host* (accessed February 3, 2016), Accession# 102151670.

22. Seyed Hossein Mousavian and Shahir Shahidsaless, *Iran and the United States: An Insider's View on the Failed Past and the Road to Peace* (New York: Bloomsbury, 2014), 32.

23. I argue, "Another point merits emphasis: The political debt that Iran owes Russia and China cannot be paid only in economic matters (energy contracts) but also in strategic vision. The possibility exists of Iran entering into a military treaty of some sort, clearing the way for China and Russia to establish military and naval bases on Iranian soil, similar to arrangements the United States has with Bahrain. However, it does not strike me as wise policy at this time. China, for instance, wants to wait till the dust settles in this latest major shift to better assess the situation." David Oualaalou, "Iran's Nuclear Agreement Divides Our Government at Its Core; Allies Won't Care," *Huffington Post*, August 3, 2015, http://www.huffingtonpost.com/david-oualaalou/irans-nuclear-agreement-divides-our-government-at-its-core-allies-wont-care_b_7922028.html.

24. "Armed Forces and Government Spending," *Iran Defence and Security Report*, January 2010, 39–44, *Business Source Complete*, EBSCO*host* (accessed February 5, 2016), Accession# 47123182 (2).

25. Alireza Nader, "The Revolutionary Guards," United States Institute of Peace: The Iran Premier, 2010, updated August 2015, http://iranprimer.usip.org/resource/revolutionary-guards.

26. "Armed Forces and Government Spending," *Iran Defence and Security Report*, January 2010, 39–44, *Business Source Complete*, EBSCO*host* (accessed February 5, 2016), Accession# 47123182 (2).

27. Anthony H. Cordesman, "The Conventional Military," United States Institute of Peace, August 2015, http://iranprimer.usip.org/resource/conventional-military.

28. Ibid.

29. Carol D. Leonnig, "Iran Held Liable in Khobar Attack," *Washington Post*, December 23, 2006, http://www.washingtonpost.com/wp-dyn/content/article/2006/12/22/AR2006122200455.html.

30. Masoud Kazemzadeh and Gabriel Emile Eid, "An Analysis of the Assassination of the Lebanese Hezbollah Commander Imad Mughniyiah: Hypotheses and Consequences," *American Foreign Policy Interests* 30, no. 6 (December 2008): 399–413. doi:10.1080/10803920802569324.

31. Shireen T. Hunter, *Iran and the World: Continuity in a Revolutionary Decade* (Bloomington: Indiana University Press, 1990), 176.

32. Oren Dorell, "Iranian Support for Yemen's Houthis Goes Back Years," *USA Today*, April 20, 2015, http://www.usatoday.com/story/news/world/2015/04/20/iran-support-for-yemen-houthis-goes-back-years/26095101/.

33. Abdel Aziz Aluwaisheg, "Iran's Meddling and US Neglect," *Arab News*, March 13, 2015, http://www.arabnews.com/columns/news/717546.

34. Henry Rome, "Elite Hezbollah Fighters Are Spearheading Battle in Syria, IDF Commander Warns," *Jerusalem Post*, October 25, 2013, http://www.jpost.com/Middle-East/Elite-Hezbollah-fighters-are-spearheading-battle-in-Syria-IDF-commander-warns-329707.

35. David Oualaalou, "Can Negotiations over Iran's Controversial Nuclear Program Defy Low Expectations?" *Waco Tribune-Herald*, November 9, 2014, http://www.wacotrib .com/opinion/columns/board_of_contributors/david-oualaalou-board-of-contributors-can -negotiations-over-iran-s/article_df375c32-448d-5b99-be03-edd958ba7696.html.

36. Tamir Eshel, "Iran Introduces a Locally Produced Kornet-Ecopy," *Defense Update*, July 7, 2012, http://defense-update.com/20120707_iran-introduces-a-locally-produced-kornet-e-copy .html.

37. "Iran to Boost Missile Arsenal after US Slaps New Ban," *Al-Jazeera*, March 26, 2016, http://www.aljazeera.com/news/2016/03/iran-boost-missile-arsenal-slaps-ban-160326184616532 .html.

38. "Iran Builds Own Aerial Drones with Strike Capabilities," *Sputnik News: Military & Intelligence*, August 2, 2010, http://sputniknews.com/military/20100208/157809895 .html#ixzz43b5kxFbe.

39. Aryeh Savir, "Iran Unveils New Missile, Radar Systems," *United with Israel*, September 3, 2014, http://unitedwithisrael.org/iran-unveils-new-missile-radar-systems/.

40. "Background Note: Iran." *Background Notes on Countries of the World: Iran*, June 2004, *MAS Ultra—School Edition*, EBSCOhost (accessed February 6, 2016), Accession# 15305054 (1).

41. Denise Youngblood Coleman, "Iran 2015 Country Review," *Iran Country Review*, July 2015, 1–479, *Business Source Complete*, EBSCOhost (accessed February 6, 2016), Accession# 102151670 (218).

42. Barbara Ann Flanagan-Rieffer, *Evolving Iran: An Introduction to Politics and Problems in the Islamic Republic* (Washington, DC: Georgetown University Press, 2013), 155.

43. Maksud Djavadov, "Iran-Russia Relations in Historical Perspective," *Crescent International of the Institute of Contemporary Islamic Thought (ICIT)*, March 2010, http://www .crescent-online.net/2010/03/iran-russia-relations-in-historical-perspective-maksud-djavadov -2533-articles.html.

44. Christopher De Ballaigue, *Patriot of Persia: Muhammad Mossadegh and a Tragic Anglo-American Coup* (New York: HarperCollins, 2012).

45. Nozhan Etezadosaltaneh, "Lessons from the Tudeh Party for the Middle East," *International Policy Digest*, February 19, 2016, http://intpolicydigest.org/2016/02/19/lessons-from-the -tudeh-party-for-the-middle-east/.

46. "New Report Details Massive Lost US Job Opportunities Due to Iran Sanctions," National Iranian American Council, July 14, 2014, http://www.niacouncil.org/report-iran -sanctions-cost-us-economy-175-billion/.

47. Janet Larsen, "Iran's Birth Rate Plummeting at Record Pace: Success Provides a Model for Other Developing Countries," Minnesotans for Sustainability, December 28, 2001, http:// www.mnforsustain.org/iran_model_of_reducing_fertility.htm.

48. "Iran-Iraq War (1980–19880)," Global Security Organization, November 7, 2011, http:// www.globalsecurity.org/military/world/war/iran-iraq.htm.

49. Patrick Clawson, "US Sanctions," United States Institute of Peace, 2010, http://iranprimer .usip.org/resource/us-sanctions.

50. Ibid.

51. "Iran: Crude Price Pegged at dlrs 39.6 a Barrel under Next Year's Budget," *Payvand Iran News*, January 27, 2008, http://www.payvand.com/news/08/jan/1250.html.

52. I argue, "All this comes on the heels of a sharp drop in oil prices to below $33 a barrel, something we see reflected whenever we pump gasoline to our heart's content or read about fiscal jitters in states that rely heavily on oil revenue to fund government operations and services. Now that sanctions are officially lifted, oil prices will continue to decline further. Yes, economic opportunities await Iran but they won't come too fast. Iran's economy consists of a series of knots that will take time to loosen." David Oualaalou, "Weighing Opportunity, Risk

in Iran, Saudi Arabia and Global Economy," *Waco Tribune-Herald*, January 28, 2016, http://www.wacotrib.com/opinion/columns/board_of_contributors/david-oualaalou-board-of-contributors-weighing-opportunity-risk-in-iran/article_a1dc39ff-c60f-5b89-a853-c8de15357450.html.

53. Ali Khaksari, Timothy Jeonglyeol Lee, and Choong-Ki Lee, "Religious Perceptions and Hegemony on Tourism Development: The Case of the Islamic Republic of Iran," *International Journal of Tourism Research* 16, no. 1 (January 2014): 97–103. *Business Source Complete*, EBSCO*host* (accessed February 10, 2016), Accession# 93449689.

54. Akram Gheibi Hashemabadi, "Ranking and Determination of Iran Status in the Middle East Region on the Development Indicator of Tourism during the Development Plan with Topsis Method," *Journal of Economic Development, Management, IT, Finance and Marketing* 7, no. 2 (September 2015): 24–36. *Business Source Complete*, EBSCO*host* (accessed February 10, 2016), Accession# 11795914 (1).

55. Emily Brennan, "Tour Iran? Operators Hope So," *New York Times*, August 29, 2013, http://www.nytimes.com/2013/09/01/travel/tour-iran-operators-hope-so.html?_r=0.

56. Martin Ferguson, "Iran Persian Promise," *Buying Business Travel* no. 77 (November 2015): 94–97. *Small Business Reference Center*, EBSCO*host* (accessed February 10, 2016), Accession# 112291030.

57. Hashem Kalantari and Golnar Motevalli, "Iran Oil Minister Says Output to Rise a Week after Sanctions," *Bloomberg Business*, August 2, 2015, http://www.bloomberg.com/news/articles/2015-08-02/iran-s-oil-minister-says-output-to-rise-one-week-after-sanctions.

58. "Enerdata: Iran Energy Report," *Global Energy Market Research: Iran* (January 2012): 1–27, *Business Source Complete*, EBSCO*host* (accessed February 10, 2016), Accession# 70374177.

59. Howard Amos, "Economic Woes Ripple across Russia's Heartland," *International Business Times*, March 11, 2016, http://www.msn.com/en-us/money/markets/economic-woes-ripple-across-russias-heartland/ar-AAgDb8Z?li=BBnbfcL&ocid=mailsignout.

60. "Europe Must Impose Tough Oil Sanctions on Iran: US," *European Business*, June 2010, http://www.eubusiness.com/news-eu/iran-nuclear-us.5db.

61. "Nuclear Power in Armenia," World Nuclear Association, October 2015, http://www.world-nuclear.org/information-library/country-profiles/countries-a-f/armenia.aspx.

62. Joel Schectman, "U.S. Imposes Ballistic Missile Sanctions on Iran after Prisoner Release," *The Star Online*, January 18, 2016, http://www.thestar.com.my/news/world/2016/01/18/us-imposes-ballistic-missile-sanctions-on-iran-after-prisoner-release/.

4. Political Landscape of the Middle East

1. Jon B. Alterman, "The Middle East's Centenarian," Center for Strategic and International Studies, May 12, 2016, https://www.csis.org/analysis/middle-easts-centenarian.

2. David Crist, *The Twilight War: The Secret History of America's Thirty-Year Conflict with Iran* (New York: Penguin, 2012), 442.

3. Gao Zugui, "Impact of the Changing Situation in the Middle East to the U.S. Strategy," China Institute of International Studies, July 2011, http://www.ciis.org.cn/english/2011–07/04/content_4310022.htm.

4. Noman Benotman and Hayden Pirkle, "The Middle East's Changing Political Landscape: A New Quilliam Strategic Assessment of the Positional Evolution of Hizb al-Nahda and the Muslim Brotherhood," *Quilliam*, May 15, 2013, http://www.quilliamfoundation.org/press/the-middle-easts-changing-political-landscape-a-new-quilliam-strategic-assessment-of-the-positional-evolution-of-hizb-al-nahda-and-the-muslim-brotherhood-2/.

5. Suleiman K. Kassicieh and Jamal R. Nassar, "Political Risk in the Gulf: The Impact of the Iran-Iraq War on Governments and Multinational Corporations," *California Management Review* 28, no. 2 (1986): 69–86, *Business Source Complete*, EBSCO*host* (accessed May 21, 2016).

6. Ibid.

7. Murray Williamson and Kevin M. Woods, *The Iran-Iraq War: A Military and Strategic History* (Cambridge, UK: Cambridge University Press, 2014), 6.

8. "The Iran-Iraq War: Serving American Interests," Iran Chamber Society, May 19, 2016, http://www.iranchamber.com/history/articles/iran_iraq_war_american_interest.php.

9. Stephen C. Pelletiere, *The Iran-Iraq War: Chaos in a Vacuum* (New York: Praeger, 1992), 6.

10. Kurt Burch, "Iran-Iraq War," *Salem Press Encyclopedia*, January 2015, *Research Starters*, EBSCO*host* (accessed April 18, 2016).

11. "The Iran-Iraq War: Serving American Interests," Iran Chamber Society, May 19, 2016, http://www.iranchamber.com/history/articles/iran_iraq_war_american_interest.php.

12. Glenn Frankel, "How Saddam Built His War Machine—with Western Help," *Washington Post Foreign Service*, September 17, 1990, A01, http://www.washingtonpost.com/wp-srv/inatl/longterm/iraq/stories/wartech091790.htm.

13. Norm Dixon, "How the US Armed Saddam with Chemical Weapons," *Green Left Weekly*, September 2, 2002, http://rense.com/general28/chmm.htm.

14. "The Iran-Iraq War: Serving American Interests," Iran Chamber Society, May 19, 2016, http://www.iranchamber.com/history/articles/iran_iraq_war_american_interest.php.

15. Jeffrey Fleishman, "Germans Were Chief Weapons Supplier to Iraq, Paper Reports," *Los Angeles Times*, December 18, 2002, http://articles.latimes.com/2002/dec/18/world/fg-german18.

16. Behnam Ben Taleblu, "The Long Shadow of the Iran-Iraq War," *National Interest*, October 23, 2014, http://nationalinterest.org/feature/the-long-shadow-the-iran-iraq-war-11535.

17. Shahram Chubin and Charles Tripp, *Iran and Iran at War* (Boulder, CO: Westview Press, 1988), 37.

18. Behnam Ben Taleblu, "The Long Shadow of the Iran-Iraq War," *National Interest*, October 23, 2014, http://nationalinterest.org/feature/the-long-shadow-the-iran-iraq-war-11535.

19. Ibid.

20. Ibid.

21. Vali Nasr, *The Shia Revival: How Conflicts within Islam Will Shape the Future* (New York: W.W. Norton & Company, 2006), 241.

22. Kurt Burch, "Iran-Iraq War," *Salem Press Encyclopedia*, January 2015, *Research Starters*, EBSCO*host* (accessed April 18, 2016).

23. Paul Kerr, "IAEA: More Questions on Iran Nuclear Program," Arms Control Association, July 1, 2005, http://www.armscontrol.org/act/2005_07-08/IAEA_Iran.

24. Adam Kredo, "Iran Violates Past Nuclear Promises on Eve of Deal," *Washington Free Beacon*, July 2, 2015, http://freebeacon.com/national-security/iran-violates-past-nuclear-promises-on-eve-of-deal/#.

25. News Agency, Iran, "Iran-Nuclear-Security," *Arabia 2000*, March 9, 2006, *Newspaper Source*, EBSCO*host* (accessed January 13, 2016).

26. Kelsey Davenport, "History of Official Proposals on the Iranian Nuclear Issue," Arms Control Association, January 2014, http://www.armscontrol.org/factsheets/Iran_Nuclear_Proposals.

27. Steven Emerson and Joel Himelfarb, "Would Iran Provide a Nuclear Weapon to Terrorists?" *In Focus* 3, no. 4 (Winter 2009), http://www.jewishpolicycenter.org/1532/iran-nuclear-weapon-to-terrorists.

28. News Agency, Iran, "Iran-Nuclear-Security," *Arabia 2000*, March 9, 2006, *Newspaper Source*, EBSCO*host* (accessed January 13, 2016).

29. Farideh Farhi, "Ahmadinejad's Nuclear Folly," *Middle East Report*, no. 252 (Fall 2009), http://www.merip.org/mer/mer252/ahmadinejads-nuclear-folly.

30. Ibid.

31. "Economic Jihad," *Economist*, June 23, 2011, http://www.economist.com/node/18867440.

32. David E. Sanger and Nazila Fathi, "Iran Test-Fires Missile with 1,200-Mile Range," *New York Times*, May 20, 2009, http://www.nytimes.com/2009/05/21/world/middleeast/21iran.html?_r=2.

33. Abbas Milani, "Obama's Existential Challenge to Ahmadinejad," *Washington Quarterly* 32, no. 2 (Spring 2009): 63, EBSCO*host* (accessed February 21, 2016), Accession# 43429688.

34. "Iran Ready for Nuclear Agreement," *Al-Jazeera Middle East*, October 29, 2009, http://www.aljazeera.com/news/middleeast/2009/10/2009102984633409448.html.

35. Hanif Zarrabi Kashani, "Iran Headlines: P5+1 Ministers in Vienna, Ahmadinejad, and Rafsanjani," *Brookings Institute*, July 11, 2014, http://www.brookings.edu/blogs/markaz/posts/headlines-2014/07/11-ministers-join-iran-nuclear-talks-vienna.

36. Hashem Kalantari (reporter); Robin Pomeroy (writer); Jon Hemming (editor), "Ahmadinejad Aide Says Iran Not Ready to Talk Nuclear," *Reuters*, October 31, 2010, http://www.reuters.com/article/us-iran-nuclear-talks-idUSTRE69U0EN20101031.

37. Clovis Maksoud, "Rhouhani's Election Win in Iran Promises to Ease Tension," *Al Monitor: The Pulse of the Middle East*, June 16, 2013, http://www.al-monitor.com/pulse/originals/2013/06/hassan-rouhani-iran-election-israel-nuclear.html#ixzz40gTJXYPy.

38. Nasser Saghafi-Ameri, "The Future of the NPT in the Light of Iran's Nuclear Dossier," Washington Conference Paper, December 2013, https://pugwashconferences.files.wordpress.com/2013/12/200604_tehran_iran_ameri_paper.pdf.

39. George Bunn and John B. Rhinelander, "NPT Withdrawal: Time for the Security Council to Step In," *Arms Control Association*, May 1, 2005, http://www.armscontrol.org/act/2005_05/Bunn_Rhinelander.

40. Alexandre Mansourov, "North Korea: Enduring Short-Term Pain for Long-Term Gain," The U.S.-Korea Institute at the Paul H. Nitze School of Advanced International Studies, July 12, 2013, http://38north.org/2013/07/amansourov071213/.

41. Benjamin Weinthal and Emanuele Ottolenghi, "Iran Made Illegal Purchases of Nuclear Weapons Technology Last Month," *Weekly Standard*, July 10, 2015, http://www.msn.com/en-us/news/world/iran-leader-withholds-verdict-on-nuclear-deal-vows-anti-us-policies/ar-AAd9SIY?ocid=mailsignout.

42. Bill Gertz, "Nuclear Deal Silent on Iran's Parchin Military Plant, Bushehr," *Washington Free Beacon*, July 14, 2015, http://freebeacon.com/national-security/nuclear-deal-silent-on-irans-parchin-military-plant-bushehr/.

43. Nasser Saghafi-Ameri, "The Future of the NPT in the Light of Iran's Nuclear Dossier," Washington Conference Paper, December 2013, https://pugwashconferences.files.wordpress.com/2013/12/200604_tehran_iran_ameri_paper.pdf.

44. Babak Dehghanpisheh and Bozorgmehr Sharafedin (edited by William Maclean and Robert Birsel), "Iran Leader Withholds Verdict on Nuclear Deal, Vows Anti-US Policies," *Reuters*, July 18, 2015, http://www.msn.com/en-us/news/world/iran-leader-withholds-verdict-on-nuclear-deal-vows-anti-us-policies/ar-AAd9SIY?ocid=mailsignou.

45. "China Envoy: Nuclear Energy Iran's Legitimate Right," *Arabia 2000*, October 2, 2007, *Newspaper Source*, EBSCO*host* (accessed January 23, 2016).

46. Alvin Cheng and Hin Lim, "Middle East and China's 'Belt and Road': Xi Jinping's 2016 State Visits to Saudi Arabia, Egypt and Iran," *Eurasia Review*, January 30, 2016, http://www.eurasiareview.com/30012016-middle-east-and-chinas-belt-and-road-xi-jinpings-2016-state-visits-to-saudi-arabia-egypt-and-iran-analysis/.

47. I argue, "Make no mistake: A high-level visit to Iran from the emerging global power portends significant changes to the geopolitical landscape—and not only in the Middle East. Admiral Jianguo's overtures stamp China's stated desire to deepen its ties with Iran. The question the United States should now be pondering: Is China's move to increase ties with Iran merely an extension of its economic pre-eminence—or is it one that heralds an inevitable expansion of its military power well beyond Iran? The answer is yes—and yes again." David Oualaalou, "Strange Bedfellows of China, Iran Could Mean Bad News for United States," *Waco Tribune-Herald*, November 1, 2015, http://www.wacotrib.com/opinion/columns/board _of_contributors/david-oualaalou-board-of-contributors-strange-bedfellows-of-china-iran /article_6ce056cf-e4e5-5324-becc-b8209067f2fe.html.

48. David Oualaalou, "China's Nuclear Upgrade Raises Concerns in the West," *Huffington Post*, March 2, 2015, http://www.huffingtonpost.com/david-oualaalou/chinas-nuclear-upgrade -ra_b_6395896.html.

49. Bill Hayton, *The South China Sea: The Struggle for Power in Asia* (New Haven, CT: Yale University Press, 2015).

50. Katherine Murphy, "Australia Confirms It Will Join China's Asian Infrastructure Investment Bank," *Guardian*, March 28, 2015, http://www.theguardian.com/business/2015 /mar/29/australia-confirms-it-will-join-chinas-asian-infrastructure-investment-bank.

51. Mimi Lau, "China Mounts Third Hypersonic 'Wu-14' Missile Test, US Report Says," *South China Morning Post*, December 6, 2014, http://www.scmp.com/news/china/article /1656748/china-mounts-third-hypersonic-wu-14-missile-test-us-report-says.

52. Vikas Shukla, "Russia Tested Hypersonic Nuclear Missile in February, but Failed," *Jane* magazine reported in *Value Walk*, June 26, 2015, http://www.valuewalk.com/2015/06/russia -tested-hypersonic-missile/.

53. I argue, "For all the saber rattling in the Pacific and financial brinkmanship with the United States, China strikes me as pragmatic, even disciplined, in not wanting to throw into chaos a world system that, after all, has allowed it to acquire wealth, power, influence and status. Yet China is slowly exhibiting its ability to challenge the United States, at least regionally. China will base its foreign policy in coming years on strategic decisions, not short-term gains." David Oualaalou, "Scattered U.S. Policies Only Fueling China's Strength Economically, Militarily and Politically," *Huffington Post*, July 10, 2015, http://www.huffingtonpost.com/david -oualaalou/scattershot-us-policies-o_b_7763354.html.

54. I argue, "Russia this month unexpectedly shelved delivery of its S-300 air defense system to Iran and replaced it with one that is more sophisticated, more powerful. The Antey-2500 defense system was developed in the 1980s and can engage missiles traveling at 4,500 meters per second, with a range of 1,500 miles. This strengthens Iran's position in nuclear talks with the United States. A recent visit by Russian defense minister Sergei Shoigu to Iran even suggests the possibility of military cooperation between the two countries." David Oualaalou, "Russia Seizes Leadership Role from US in Middle East Upheaval," *Waco Tribune-Herald*, October 9, 2015, http://www.wacotrib.com/opinion/columns/board_of_contributors/david -oualaalou-board-of-contributors-russia-seizes-leadership-role-from/article_dec4be5e-2367 -57a1-b0cd-ed1933a739d0.html.

55. David Oualaalou, "Russia Outsmarts the U.S. in Syria: A New Geopolitical Outlook," *Huffington Post*, October 2, 2015, http://www.huffingtonpost.com/david-oualaalou/russia --us_b_8273562.html.

56. "Moscow Still for a Negotiated Settlement to Iran Nuclear File," *Arabia 2000*, September 13, 2007, *Newspaper Source*, EBSCO*host* (accessed January 23, 2016).

57. "Russia Hopes Experience of Agreement on Iran's Nuclear Program Will Be Useful for Resolving Other Crisis Situations in World—Russian Permanent Envoy to UN (Part 2),"

Interfax: Russia & CIS Military Newswire, July 20, 2015, 1, *Regional Business News*, EBSCO*host* (accessed January 23, 2016).

58. Nick Bryant, "The Decline of US Power?" *BBC News*, July 10, 2015, http://www.bbc.com /news/world-us-canada-33440287.

59. I previously argued, "A few decades ago, it would have been inconceivable to think that China or Russia, for that matter, would be selling advanced missile technology to a staunch US ally such as Saudi Arabia." David Oualaalou, *The Ambiguous Foreign Policy of the United States toward the Muslim World: More than a Handshake* (Lanham, MD: Rowman & Littlefield, 2016).

60. Laura K. Egendorf, *Iran: Opposing Viewpoints* (Farmington Hill, MI: Greenhaven, 2006), 116.

61. Adam Kredo, "Russian Warships Dock in Iran for War Training," *Washington Free Beacon*, August 10, 2015, http://freebeacon.com/national-security/russian-warships-dock-in-iran -for-war-training/.

62. Gabriela Baczynska, Tom Perry, Laila Bassam, and Phil Stewart, "Russian Troops Join Combat in Syria," *Reuters*, September 9, 2015, http://www.reuters.com/article/us-mideast -crisis-syria-exclusive-idUSKCN0R91H720150909.

63. Mark Galeotti and Jonathan Spyer, "Russia to Defend Core Syrian Government Areas," *Jane's Intelligence Review*, September 22, 2015, http://www.janes.com/article/54732/russia -to-defend-core-syrian-government-areas.

64. Alastair Crooke, "Russia Isn't Really Withdrawing from Syria," *Huffington Post*, March 17, 2016, http://www.huffingtonpost.com/alastair-crooke/russia-withdraw-syria_b_9487262 .html.

65. Burak Bekdil, "The European Union Caves to Turkey's Blackmail," *Middle East Forum*, March 15, 2016, http://www.meforum.org/5910/turkey-eu-blackmail.

66. David Oualaalou, "NATO's Expansion in Membership Only Provoking Russia," *Waco Tribune-Herald*, December 31, 2015, http://www.wacotrib.com/opinion/columns/board_of _contributors/david-oualaalou-board-of-contributors-nato-s-expansion-in-membership /article_a61c959d-534b-5895-8d33-2c2a51bee93b.html.

67. Adam Schreck and Lee Keath, "In Arab World, Worries that Deal Will Boost Iran's Power," *Yahoo News*, July 15, 2015, https://www.yahoo.com/news/arab-world-worries-deal -boost-irans-power-064114480.html.

68. Jeff Dunetz, "Saudi Prince Threatens 'Military Action without American Support' against Iran," Media Research Center, July 17, 2015, http://www.mrctv.org/blog/saudi-prince -even-after-deal-military-action-against-iran-still-table-or-without-us#.nw8trn:5qPz.

69. Ali Akbar Dareini, "Iran Transport Minister: Agreement to Buy 114 Airbus Planes," *Associated Press*, January 16, 2016, http://portal.tds.net/news/read/category/world/article /the_associated_press-iran_transport_minister_agreement_to_buy_114_airbu-ap.

70. Alireza Noori, "Russia and Iran's Nuclear Dossier in Rouhani's Tenure: The Need for a Change," *Iran Review Analysis*, July 4, 2013, http://www.iranreview.org/content/Documents /Russia-and-Iran-s-Nuclear-Dossier-in-Rouhani-s-Tenure-The-Need-for-a-Change.htm.

71. Krysta Wise, "Islamic Revolution of 1979: The Downfall of American-Iranian Relations," *Legacy* 11, no. 1, article 2 (2012), http://opensiuc.lib.siu.edu/cgi/viewcontent.cgi?article=1008& context=legacy&sei-redir=1&referer=http%3A%2F%2Fwww.bing.com%2Fsearch%3Fq%3DUS -Iran%2Brelations%2Bbefore%2Bthe%2BRevolutions%2Bof%2B1979%26form%3DPRTOS1% 26mkt%3Den-us%26refig%3D272b7c863a2c4dd0a5f08026d9ecff5e#search=%22US-Iran%20 relations%20before%20Revolutions%201979%22.

72. Stephen Kinzer, *All the Shah's Men: An American Coup and the Roots of Middle East Terror* (Hoboken, NJ: John Wiley and Sons, 2008).

73. Krysta Wise, "Islamic Revolution of 1979: The Downfall of American-Iranian Relations," *Legacy* 11, no. 1, article 2 (2012), http://opensiuc.lib.siu.edu/cgi/viewcontent.cgi?article=1008& context=legacy&sei-redir=1&referer=http%3A%2F%2Fwww.bing.com%2Fsearch%3Fq%3DUS -Iran%2Brelations%2Bbefore%2Bthe%2BRevolutions%2Bof%2B1979%26form%3DPRTOS1% 26mkt%3Den-us%26refig%3D272b7c863a2c4ddoa5fo8026d9ecff5e#search=%22US-Iran%20 relations%20before%20Revolutions%201979%22.

74. David Oualaalou, *The Ambiguous Foreign Policy of the United States toward the Muslim World: More than a Handshake* (Lanham, MD: Lexington Books, 2016).

75. Thierry Brun, "Resurgence of Popular Agitation in Iran," *Le Monde Diplomatique*, July 17–18, 1978.

76. Kenneth M. Pollack, *The Persian Puzzle: The Conflict between Iran and America* (New York: Random House, 2005).

77. Nader Entessar, "US Foreign Policy and Iran: American–Iranian Relations since the Islamic Revolution," *Iranian Studies* 46, no. 2 (March 2013): 321–23, *Academic Search Complete*, EBSCO*host* (accessed June 2, 2016).

78. Donette Murray, *US Foreign Policy and Iran: American Iranian Relations since the Islamic Revolution* (London and New York: Routledge, 2010), 6.

79. Gary Sick, "The Carter Administration," United States Institute of Peace, http:// iranprimer.usip.org/resource/carter-administration-0.

80. Malcolm Byrne, *IRAN-CONTRA: Reagan's Scandal and the Unchecked Abuse of Presidential Power* (Lawrence: University Press of Kansas, 2014).

81. Richard N. Haass, "The George H. W. Bush Administration," United States Institute of Peace, http://iranprimer.usip.org/resource/carter-administration-0.

82. Peter W. Rodman, "Mullah Moola: The U.S. Is Trying to Maintain a Firm Policy toward Iran, but European Loans and Trade Are Undermining It," *National Review* 46, no. 21 (November 7, 1994): 66. *MasterFILE Premier*, EBSCO*host* (accessed June 7, 2016).

83. "Obama's Speech in Cairo," *New York Times*, June 4, 2009, http://www.nytimes .com/2009/06/04/us/politics/04°bama.text.html?_r=0.

84. Michael Crowley, Aryn Baker, and Zeke Miller, "Obama's Iran Gamble," *Time International (South Pacific Edition)* 182, no. 24 (2013): 30, *MasterFILE Premier*, EBSCO*host* (accessed June 8, 2016).

85. Ray Takeyh, "What Really Happened in Iran," *Foreign Affairs* 93, no. 4 (2014): 2–12, *Business Source Complete*.

86. Roland Flamini, "Iran's Nuclear Program Has a Long History," *World Politics Review*, February 6, 2007, http://www.worldpoliticsreview.com/articles/524/irans-nuclear-program-has -a-long-history.

87. Ibid.

88. "Statement by H. E. Dr. M. Javad Zarif Permanent Representative of the Islamic Republic of Iran before the Security Council," United Nations Security Council, December 23, 2006, http://www.payvand.com/news/06/aug/1001.html.

89. Leonard Weiss, "Israel's Future and Iran's Nuclear Program," *Middle East Policy* 16, no. 3 (Fall 2009): 79–88, *Academic Search Complete*, EBSCO*host* (accessed June 21, 2016).

90. "William Beeman on Iran's Election," *New York Times*, June 16, 2005, http://www .nytimes.com/cfr/international/slot2_061605.html?_r=1&pagewanted=print&oref=slogin.

91. Duyeon Kim and Ariane Tabatabai, "Should South Korea Be Iran's Next Nuclear Energy Partner?" *Bulletin of the Atomic Scientists*, May 31, 2016, http://thebulletin.org/should -south-korea-be-irans-next-nuclear-energy-partner9495.

92. Ibid.

93. Robert Tait, "Iran State TV Poll Reveals Iranians Want Nuclear Programme Stopped," *Telegraph (London)*, July 9, 2012, http://www.telegraph.co.uk/news/worldnews/middleeast /iran/9379493/Iran-state-TV-poll-reveals-Iranians-want-nuclear-programme-stopped.html.

94. Saeed Kamali Dehghan, "Ex-Revolutionary Guards General Reveals Dissent within Elite Iranian Force," *Guardian (London)*, July 12, 2012, https://www.theguardian.com/world /2012/jul/12/khamenei-accused-blood-on-hands.

95. James M. Dorsey, "UAE Toughens Stance over Iran's Nuclear Ambitions," *World Politics Review*, July 13, 2010, http://www.worldpoliticsreview.com/articles/6019/uae-toughens-stance -over-irans-nuclear-ambitions.

96. Rick Gladstone, "Bahrain's Sunni Rulers Revoke Citizenship of Top Shiite Cleric," *New York Times*, June 20, 2016, http://www.nytimes.com/2016/06/21/world/middleeast/bahrains -sunni-rulers-revoke-citizenship-of-top-shiite-cleric.html?_r=0.

97. Ian Black and Simon Tisdall, "Saudi Arabia Urges US Attack on Iran to Stop Nuclear Programme," *Guardian (London)*, November 28, 2010, https://www.theguardian.com/world /2010/nov/28/us-embassy-cables-saudis-iran.

98. Simon Mabon, *Saudi Arabia and Iran: Power and Rivalry in the Middle East* (New York and London: I. B. Tauris, 2015), 218.

99. Omar Fahmy and Asma Alsharif, "Russia Signs Deal with Egypt to Build First Nuclear Power Plant in the Country," *Haaretz (Middle East News)*, November 19, 2015, http://www .haaretz.com/middle-east-news/1.687253.

100. "Jordan to Build Nuclear Research Reactor," *Physics Org*, August 20, 2013, http://phys .org/news/2013–08-jordan-nuclear-reactor.html.

101. Tsuyoshi Inajima, "Turkey, Japan in Exclusive Talks for Nuclear Plant," *Bloomberg Technology*, December 23, 2010, http://www.bloomberg.com/news/articles/2010–12–23/turkey -in-exclusive-talks-with-japan-to-build-nuclear-plant-yildiz-says.

102. "Official: Iran 'Not Likely' to Have Nuclear Capabilities by 2010," *China View*, September 16, 2008, http://news.xinhuanet.com/english/2008–09/16/content_10024502.htm.

103. Leonard Weiss, "Israel's Future and Iran's Nuclear Program," *Middle East Policy* 16, no. 3 (Fall 2009): 79–88, *Academic Search Complete*, EBSCO*host* (accessed June 21, 2016).

104. Shlomo Ben-Ami, "Iran's Nuclear Grass Eaters," Project Syndicate, April 4, 2012, https://www.project-syndicate.org/commentary/iran-s-nuclear-grass-eaters?barrier=true.

105. "Germany Could Accept Nuclear Enrichment in Iran," *Reuters*, June 6, 2006, http:// www.chinadaily.com.cn/world/2006–06/29/content_628762.htm.

106. David Oualaalou, "Weighing Opportunity, Risk in Iran, Saudi Arabia and Global Economy," *Waco Tribune-Herald*, January 28, 2016, http://www.wacotrib.com/opinion/columns /board_of_contributors/david-oualaalou-board-of-contributors-weighing-opportunity-risk -in-iran/article_a1dc39ff-c60f-5b89-a853-c8de15357450.html.

107. "Iran's Nuclear Programme: As the Enrichment Machines Spin On," *Economist* 386, no. 8565 (February 2, 2008): 30, http://www.economist.com/node/10601584.

108. Thomas Friedman, "Obama Makes His Case on Nuclear Deal," *New York Times*, July 14, 2015, http://www.nytimes.com/2015/07/15/opinion/thomas-friedman-obama-makes-his-case -on-iran-nuclear-deal.html?_r=1.

109. David Oualaalou, "Iran's Nuclear Accord: An Opportunity, Too Much Noise, or a Strategic Mistake?" *Huffington Post*, July 24, 2015, http://www.huffingtonpost.com/david -oualaalou/irans-nuclear-accord-an-o_b_7866822.html.

110. Cathy Burke, "Russia Building Spy Base in Nicaragua," *News Max*, June 26, 2016, http:// www.newsmax.com/Newsfront/spy-base-tank-sale-russia-managua/2016/06/26/id/735694/.

111. "Japan's Ties with Iran to Grow after Nuclear Deal Finalized, Abe Says," *Japan Times*, April 23, 2015, http://www.japantimes.co.jp/news/2015/04/23/national/politics-diplomacy/japans -ties-iran-grow-final-nuclear-deal-abe-says/#.V21aqjVWBL9.

112. "Iran-Pakistan-India Gas Pipeline (IPI)," *Gulf Oil and Gas*, February 15, 2002, http:// www.gulfoilandgas.com/webpro1/projects/3dreport.asp?id=100730.

113. "Pakistan Admits Nuclear Expert Traded with Iran," *Guardian (London)*, March 10, 2005, https://www.theguardian.com/world/2005/mar/10/iran.pakistan.

5. Nuclear Arms Race in the Middle East

1. Ari Shavit, "The Bomb and the Bomber," *New York Times*, March 21, 2012, http://www
.nytimes.com/2012/03/22/opinion/the-bomb-and-the-bomber.html.

2. Steven A. Cook, "Don't Fear a Nuclear Arms Race in the Middle East," *Foreign Policy*,
April 2, 2012, http://foreignpolicy.com/2012/04/02/dont-fear-a-nuclear-arms-race-in-the-middle
-east/.

3. Chriss W. Street, "American Oil Reserves Now Top Saudi Arabia," *Breitbart (National
Security)*, April 16, 2014, http://www.breitbart.com/national-security/2014/04/16/american-oil
-reserves-now-top-saudi-arabia/.

4. Ian J. Stewart and Dominic J. Williams, "Is Saudi Arabia Trying to Get Nuclear
Weapons?" *Telegraph (London)*, May 23, 2015, http://www.telegraph.co.uk/news/worldnews
/middleeast/saudiarabia/11617339/Is-Saudi-Arabia-trying-to-get-nuclear-weapons.html.

5. Olli Heinonen and Simon Henderson, "Nuclear Kingdom: Saudi Arabia's Atomic Ambitions," Belfer Center for Science and International Affairs, March 27, 2014, http://belfercenter
.ksg.harvard.edu/publication/24061/nuclear_kingdom.html.

6. Ari Shavit, "The Bomb and the Bomber," *New York Times*, March 21, 2012, http://www
.nytimes.com/2012/03/22/opinion/the-bomb-and-the-bomber.html?_r=0.

7. Roger Boyes, "Saudi Arabia Seals Nuclear Weapons 'Deal' with Pakistan," *Times (London)*, 42, *Newspaper Source*, November 8, 2013, EBSCO*host* (accessed June 30, 2016).

8. US Government, *Chain Reaction: Avoiding a Nuclear Arms Race in the Middle East*
(Dulles, VA: Progressive Management, 2016).

9. David Oualaalou, "Growing Sunni-Shiite Rift Fuels Power Plays, Atomic Pursuits
across Middle East," *Waco Tribune-Herald*, May 31, 2015, http://www.wacotrib.com/opinion
/columns/board_of_contributors/david-oualaalou-board-of-contributors-growing-sunni
-shiite-rift-fuels/article_52a144c3–6bc8–5247-aa87-de3205cef5e3.html.

10. "Brief History of Republic of Turkey," University of Michigan, accessed July 3, 2016,
http://www.umich.edu/~turkish/links/reptr_brhist.html.

11. David Oualaalou, *The Ambiguous Foreign Policy of the United States toward the Muslim
World: More than a Handshake* (Lanham, MD: Rowman & Littlefield, 2016).

12. Sinan Ülgen, "Turkey and the Bomb," Carnegie Endowment for International Peace,
February 15, 2012, http://carnegieendowment.org/2012/02/15/turkey-and-bomb#5.

13. Aaron Stein, "B61 Nuclear Bombs in Turkey," *Eurasian Hub*, November 22, 2012, https://
eurasianhub.com/2013/05/11/b61-nuclear-bombs-in-turkey/.

14. "Nuclear Power in Turkey," World Nuclear Organization, May 2016, http://www.world
-nuclear.org/information-library/country-profiles/countries-t-z/turkey.aspx.

15. David Oualaalou, "A Possible Coup in Saudi Arabia Signals the End of US Dominance
in the Mideast," *Huffington Post*, October 20, 2015, http://www.huffingtonpost.com/david
-oualaalou/a-possible-coup-in-saudi_b_8325456.html.

16. Kathleen Moore and Brian Whitmore, "Obama, in Prague, Calls for Elimination of
Nuclear Weapons," *Radio Free Europe*, April 5, 2009, http://www.rferl.org/content/Obama
_Calls_For_Elimination_Of_Nuclear_Weapons_In_Prague_Speech/1602285.html.

17. Sinan Ülgen, "Turkey and the Bomb," Carnegie Endowment for International Peace,
February 15, 2012, http://carnegieendowment.org/2012/02/15/turkey-and-bomb#5.

18. Karl Vick, "The Middle East Nuclear Race Is Already Under Way," *Time*, March 23, 2015,
http://time.com/3751676/iran-talks-nuclear-race-middle-east/.

19. "Egypt Recovers Money from Mubarak's Swiss Bank Accounts," *Al Arabiya*, June
5, 2014, http://english.alarabiya.net/en/News/middle-east/2014/06/06/Egypt-recovers-money
-from-Mubarak-s-Swiss-bank-accounts-.html.

20. "U.S. Plans to Boost Aid to Jordan to $1 Billion per Year," *Reuters*, February 3, 2015, http://www.reuters.com/article/us-jordan-aid-idUSKBN0L72ET20150203.

21. "Jordan to Build Nuclear Research Reactor," *Physics Org*, August 20, 2013, http://phys.org/news/2013–08-jordan-nuclear-reactor.html.

22. Chen Kane, "Are Jordan's Nuclear Ambitions a Mirage?" *Bulletin of the Atomic Scientists*, December 16, 2013, http://thebulletin.org/are-jordans-nuclear-ambitions-mirage.

23. Ibid.

6. The Future of the Middle East and Its Geopolitical Outcome

1. Reva Goujan, "Quantum Geopolitics," Stratfor, July 28, 2015, http://www.financialsense.com/contributors/reva-bhalla/quantum-geopolitics.

2. Steve Hargreaves, "How Iran Could Double Its Oil Output," *CNN Money Invest*, May 21, 2012, http://money.cnn.com/2012/05/21/markets/iran_oil/index.htm.

3. Awad Mustafa, "Russia: Iran to Receive S-300 Air Defenses by Year's End," *Defense News*, May 19, 2016, http://www.defensenews.com/story/defense/international/2016/05/19/russia-iran-s-300-air-defense/84581758/.

4. Joel Wuthnow, "Are Chinese Arms about to Flood into Iran?" *National Interest*, January 13, 2016, http://nationalinterest.org/feature/are-chinese-arms-about-flood-iran-14887.

5. David Oualaalou, "A Possible Coup in Saudi Arabia Signals the End of US Dominance in the Mideast," *Huffington Post*, October 20, 2015, http://www.huffingtonpost.com/david-oualaalou/a-possible-coup-in-saudi_b_8325456.html.

6. Gregory F. Gause III, "The Saudis Can't Rein in Islamic State. They Lost Control of Global Salafism Long Ago," *Los Angeles Times*, July 19, 2016, http://www.latimes.com/opinion/op-ed/la-oe-gause-saudi-arabia-extremism-blame-20160719-snap-story.html.

7. Natalie Johnson, "German Intel Says Iran Attempting to Obtain Illegal Nuclear Material," *Washington Free Beacon*, July 8, 2016, http://freebeacon.com/national-security/german-intel-says-iran-attempting-obtain-illegal-nuclear-material/.

8. Bret Stephens, "Iran Cannot Be Contained," *Commentary* 130, no. 1 (2010): 61, *MasterFILE Premier*, EBSCO*host* (accessed June 7, 2016).

9. Dale Sprusansky, "The Saudi Arabia-Iran Divide: What Lies Ahead?" *Washington Report on Middle East Affairs* 35, no. 2 (March 2016): 56–57, *Military & Government Collection*, EBSCO*host* (accessed July 21, 2016).

10. Tom Phillips, Oliver Holmes, and Owen Bowcott, "Beijing rejects tribunal's ruling in South China Sea Case," the *Guardian*, July 12, 2016, https://www.theguardian.com/world/2016/jul/12/philippines-wins-south-china-sea-case-against-china.

11. "Mutual Defense Treaty between the United States and the Republic of the Philippines; August 30, 1951," *Yale Law School*, 2008, http://avalon.law.yale.edu/20th_century/phil001.asp.

12. David Oualaalou, "China's Nuclear Upgrade Raises Concerns in the West," *Huffington Post*, March 2, 2015, http://www.huffingtonpost.com/david-oualaalou/chinas-nuclear-upgrade-ra_b_6395896.html.

13. Agence France-Presse, "Russia's Defence Minister on Surprise Visit to Iran," *Defense News*, February 21, 2016, http://www.defensenews.com/story/defense/2016/02/21/russias-defence-minister-surprise-visit-iran/80708442/.

14. "Iran Receives Advanced Missile System from Russia," *Al-Jazeera*, April 11, 2016, http://www.aljazeera.com/news/2016/04/iran-receives-advanced-missile-system-russia-160411131008684.html.

15. Tim Daiss, "Beijing and Moscow Fed Up with Washington, Time for a New Order," *Forbes*, June 27, 2016, http://www.forbes.com/sites/timdaiss/2016/06/27/beijing-and-moscow-fed -up-with-washington-time-for-a-new-order/#1f23de9c1c1a.

16. Ibid.

17. James R. Holmes, "China's 'String of Pearls': Naval Rivalry or Entente in the Indian Ocean?" *World Politics Review*, March 1, 2016, http://www.worldpoliticsreview.com/articles /18085/china-s-string-of-pearls-naval-rivalry-or-entente-in-the-indian-ocean.

18. Ibid.

19. Michael Martina and Lesley Wroughton, "Diplomatic Win for China as ASEAN Drops Reference to Maritime Court Ruling," *Reuters*, July 25, 2016, https://www.yahoo.com/news /asean-breaks-deadlock-south-china-sea-beijing-thanks-061135765.html.

20. Maxim A. Suchkov, "Russia's Plan for the Middle East," *National Interest*, January 15, 2016, http://nationalinterest.org/feature/russias-plan-the-middle-east-14908?page=3.

21. David Oualaalou, "Putin Proves Ever-Mercurial in Russia's Battlefield Drawdown in War-Torn Syria," *Waco Tribune-Herald*, March 27, 2016, http://www.wacotrib.com/opinion /columns/board_of_contributors/david-oualaalou-board-of-contributors-putin-proves-ever -mercurial-in/article_48907b46-eabb-5574–834d-0d8031f1dbb3.html.

22. "Week 13 of Russia's Syria Campaign: Debunking the Lies," *Russia Insider*, January 4, 2016, http://russia-insider.com/en/military/week-13-russias-syria-campaign-debunking-lies /ri12040.

23. Maxim A. Suchkov, "Russia's Plan for the Middle East," *National Interest*, January 15, 2016, http://nationalinterest.org/feature/russias-plan-the-middle-east-14908?page=3.

24. Kevin A. Lees, "Nobody Knows Who'll Be in Charge after Oman's 'Founding Father,'" *National Interest*, July 31, 2016, http://nationalinterest.org/feature/nobody-knows-wholl-be -charge-after-omans-founding-father-17195.

25. Payam Mohseni, "Tipping the Balance? Implications of the Iran Nuclear Deal on Israeli Security," Belfer Center for Science and International Affairs, December 10, 2015, 28, http:// belfercenter.ksg.harvard.edu/files/Tipping%20the%20Balance%20WEB.pdf.

26. Bret Stephens, "Iran Cannot Be Contained," *Commentary* 130, no. 1 (2010): 61, *Master-FILE Premier*, EBSCOhost (accessed August 4, 2016).

27. Ibid.

28. David Oualaalou, *The Ambiguous Foreign Policy of the United States toward the Muslim World: More than a Handshake* (Lanham, MD: Rowman & Littlefield, 2016).

7. Author's Reflections

1. Jon B. Alterman, "The Middle East's Centenarian," The Center for Strategic and International Studies (CSIS), May 12, 2016, https://www.csis.org/analysis/middle-easts-centenarian.

2. Mohammad Shafiq Hamdam, "The Evolving Relationship between U.S. and Iran," *Huffington Post*, August 9, 2016, http://www.huffingtonpost.com/mohammad-shafiq-hamdam /usiran-evolving-relations_b_11383952.html.

Index

USSR, 73, 106–7, 144, 177, 179, 187
  collapse of, 10, 57, 115, 117–18
  former, 156
  United States and, 59, 68, 72, 74

Vatican, 42
Venezuela, 37, 56
Vick, Karl, 146
Vientiane, Laos, 169
Vietnam, 8, 94
Vimont, Pierre, 126
Visegrad, 170

Wahhab, Muhammad Abd al-, 22
Washington, 26–27, 35, 52, 59, 64, 84, 87, 95, 114, 119,
  136–38, 156, 160, 162–64, 172, 180–83, 192–93
  Beijing and, 190
  establishment, 17, 68–69, 90, 93, 96, 99, 100,
    110, 185–87, 191
  foreign policy, 15
  London and, 36–37
  Moscow and, 24, 105, 157, 170–71
  post, 75
  Riyadh and, 23, 157, 171, 188
  Tehran and, 98, 110, 116, 171, 189
  Turkey and, 17–18
Weinberg, David, 52

Weinthal, Benjamin, 89
Weiss, Leonard, 120, 125
Wuthnow, Joel, 158

Yemen, 22–23 69, 132, 165, 176
  anarchy in, 1, 13
  conflict in, 6–7, 13, 20, 22
  Houthi in, 3, 16, 51–52, 70, 79, 160–61
  Iraq and, 3–4, 26, 64, 68, 70, 108, 133, 149, 160,
    174, 187, 191–92
  Lebanon and, 3–4, 7, 70, 149, 174, 191–92
  Libya and, 12, 64, 67–68, 146, 174, 187
  Saudi Arabia and, 62, 102, 123, 149, 174
  Syria and, 4, 6–7, 12, 26, 28, 58, 64, 70, 98,
    102–3, 123, 133, 135, 149, 158–60, 162, 184, 187,
    191–92
  turmoil in, 2, 14–15, 21, 115
  upheaval in, 3, 6, 18–19, 97, 110, 175

Zabardast, Shalaleh, 48
Zagros Mountains, 30, 42
Zahedi, Fazlollah, 106
Zakaria, Fareed, 177
Zarif, Mohammad Javad, 53, 120
Zarrabi-Kashani, Hanif, 86
Zoroaster, 31
Zugui, Gao, 68

DAVID OUALAALOU is a global-affairs analyst and former international security analyst in Washington, DC. He is author of *The Ambiguous Foreign Policy of the United States toward the Muslim World: More than a Handshake.*

CPSIA information can be obtained
at www.ICGtesting.com
Printed in the USA
LVOW13s1601180218
567037LV00019B/114/P